Structure and Dynamics of Forests under
Logging Concession in Papua, Indonesia

Structure and Dynamics of Forests under Logging Concession in Papua, Indonesia

Dissertation

to attain the doctoral degree Doctor of Philosophy (Ph.D.)
of the Faculty of Forest Sciences and Forest Ecology,
Georg-August-Universität Göttingen

Submitted By

Yosias Gandhi

Born in Merauke, Indonesia

Göttingen, 2021

Bibliographical information held by the German National Library

The German National Library has listed this book in the Deutsche Nationalbibliografie (German national bibliography); detailed bibliographic information is available online at http://dnb.d-nb.de.

1st edition - Göttingen: Cuvillier, 2022

1st Referee : **Prof. Dr. Ralph Mitlöhner**

2nd Referee: **Prof. Dr. Renate Bürger-Arndt**

Date of oral examination: 22.06.2021

© CUVILLIER VERLAG, Göttingen, Germany 2022
  Nonnenstieg 8, 37075 Göttingen, Germany
  Telephone: +49 (0)551-54724-0
  Telefax: +49 (0)551-54724-21
  www.cuvillier.de

  ISBN 978-3-7369-7634-4
  eISBN 978-3-7369-6634-5

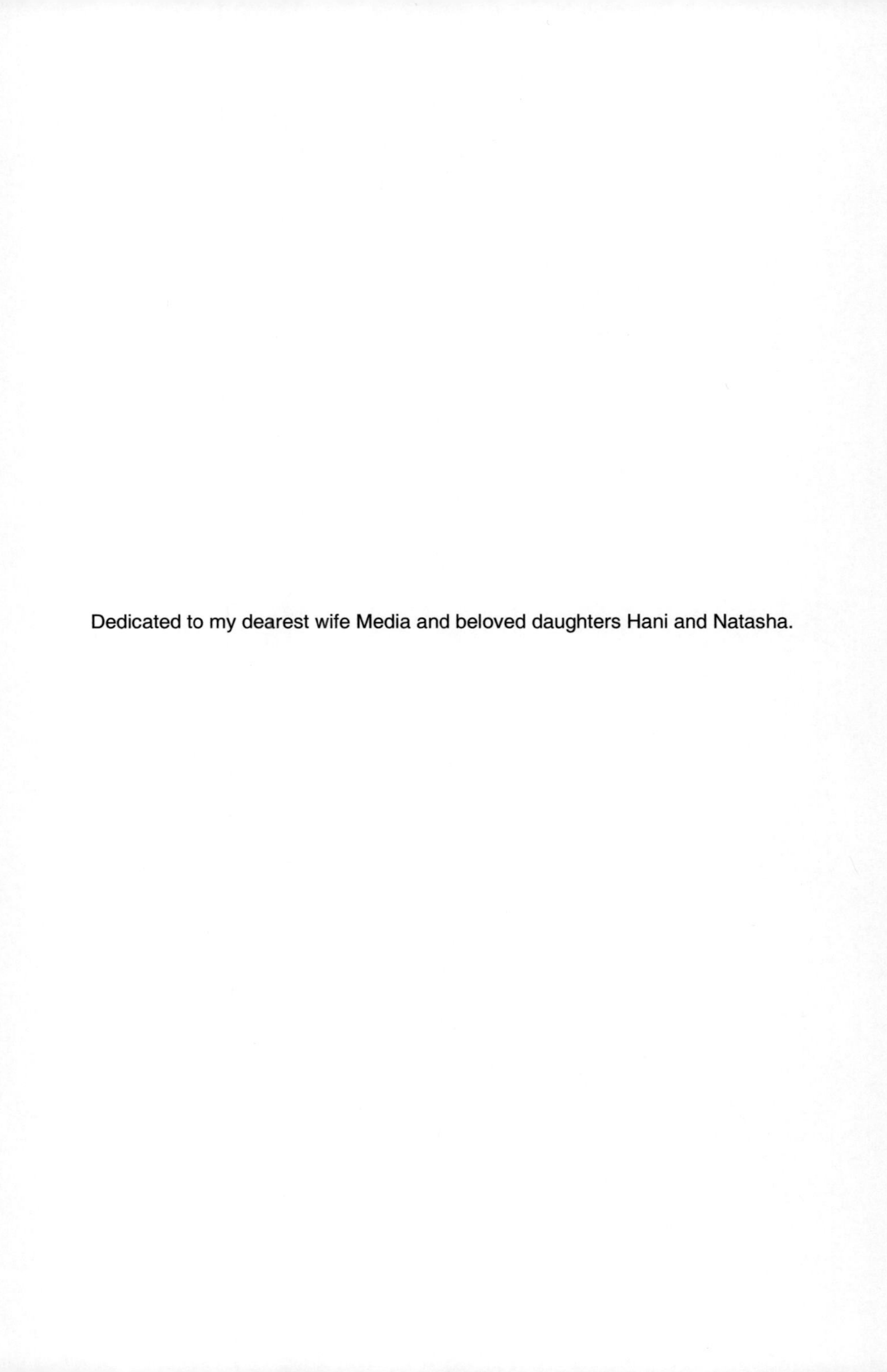

Dedicated to my dearest wife Media and beloved daughters Hani and Natasha.

Acknowledgements

My deep gratitude goes first to Prof. Dr. Ralph Mitlöhner, the Department of Tropical Silviculture and Forest Ecology. His expertise was invaluable in guiding throughout this dissertation's writing with valuable ideas and heartful support. Your insightful feedback pushed me to sharpen my thinking and brought my work to a higher level.

My sincere thanks also go to Prof. Dr. Renate Bürger-Arndt, Department of Nature Conservation and Landscape Planning, and Prof. Dr. Max Krott, Department of Forest and Nature Conservation Policy and Forest History, the University Göttingen, for having examined and reviewed my dissertation and their valuable advice.

Without the financial support of the Directorate General of Higher Education (DGHE) - Indonesian Ministry of Education, which offered me a scholarship for the Ph.D. program, this work would not have been possible. The Centre of Biodiversity and Sustainable Land Use, and the International Office, University Göttingen, are also acknowledged for providing small grants for the research. I wish to extend my special thanks to the University of Papua Manokwari, and Bupati Kabupaten Manokwari Selatan, Papua Barat, Indonesia, which financially contributed to the completion of the dissertation. My appreciation also extends to the Operational Director of the logging companies, P.T. Tunas Timber Lestari and P.T. Wapoga Mutiara Timber, in Papua, for facilitating the data collection in Tunas and Bonggo forests.

I owe deep gratitude to all colleagues at the Department of Tropical Silviculture and Forest Ecology, the University Göttingen, for their understanding, support and advice during my study period. My great thanks go to Eva Siegelkow, Reinhard Kopp, Dr. Sophie Graefe, Hombe Gouda, Afik Hardanto, and Tedy Yunanto. They all helped me in various ways, for which I am forever indebted.

I wish to express my thanks to other friends and colleagues who, in a particular way, made my stay in Göttingen, Germany, exciting and worth remembering. May also wish to the Indonesian Christian Community (PERKI-HANNA) in Göttingen and the Indonesian student association in Goettingen (PPI Goe) for making my life home.

Last but not least, I would like to express my heart-felt gratitude to my parents, my beloved wife (Media), my lovely daughters (Hani and Natasha) and all my brothers and sisters for their encouragement and assistance throughout my life. Without their support, this work has never been accomplished and my success has never come true.

Table of Contents

1. Introduction

1.1. The threat to tropical forests of insular South East Asia 1

1.2. Regeneration of natural tropical primary forest 3

1.3. Logging impacts on tropical forests 6

1.4. Sustaining tropical logged forests 8

1.5. The rationale of the study 11

2. Study sites and Methods

2.1. Site selection 16

2.2. Tunas site

2.2.1. Location 17

2.2.2. Climate 17

2.2.3. Topography and hydrology 18

2.2.4. Geology and soils 19

2.2.5. Forest history and resources 20

2.3. Bonggo site

2.3.1. Location 22

2.3.2. Climate 22

2.3.3. Topography and hydrology 23

2.3.4. Geology and soils 24

2.3.5. Forest history and resources 26

2.4. Study plot establishment, arrangement, and design 27

2.5. Data Collection

2.5.1. Measurement in subplots (compartments) 29

2.5.2. Wood sample collection and preparation for nitrogen analysis .. 31

2.6. Data analysis 35

3. Accuracy and precision of sample data

    3.1. Data accuracy 36

    3.2. Data precision 38

4. Intact primary forests of Tunas and Bonggo

    4.1. Tree family and species composition

        4.1.1. Family composition 41

        4.1.2. Species composition 48

        4.1.3. Stand composition at another site in western New Guinea 53

    4.2. Tree species diversity and similarity in the primary forest

        4.2.1. Species richness 56

        4.2.2. Species diversity indices 62

        4.2.3. Species similarity 66

        4.2.4. Tree diversity in comparisons to other sites in New Guinea 67

    4.3. Stand structure and dynamics in the primary forest

        4.3.1. Stem density, mean diameter, and basal area (BA) 70

        4.3.2. Diameter distribution and slope of population structure 73

        4.3.3. Species distribution in tree size classes 77

        4.3.4. Stand height 79

        4.3.5. Vertical structure distribution 80

        4.3.6. Diameter-height curves 83

    4.4. Seedling and sapling characteristics of the primary forest

        4.4.1. Species composition 88

        4.4.2. Species diversity 93

        4.4.3. Regeneration of canopy tree species 95

        4.4.4. Regeneration abundance 97

        4.4.5. Diameter and height of saplings 99

5. Logged forests of Tunas and Bonggo

    5.1. Tree species composition in unlogged and logged forests 105

    5.2. Tree species richness and diversity in the logged forests

        5.2.1. Species richness 119

        5.2.2. Species diversity indices 120

    5.3. Stand structure of logged forest

        5.3.1. Diameter distribution...123

        5.3.2. Species distribution in diameter size classes...........................127

    5.4. Regeneration of canopy tree species 130

6. Light requirements of tree species

    6.1. Concept 135

    6.2. Slope of tree species distribution 138

    6.3. Stable N isotope ratio (δ^{15}N) and total N in tree tissue 149

    6.4. Diameter distribution, stable N isotope ratio (δ^{15}N) and total N 156

    6.5. Grouping of tree species light requirements 160

7. Silvicultural approaches and further implications

    7.1. Silviculture: General definitions and implications................................169

    7.2. Silviculture: Existing regulations 170

    7.3. Implications of species performance for silviculture approach 171

8. Summary..175

References 188

Appendices 214

List of Tables

Table 3.1.   Mean basal area and standard error (in absolute and percentage terms) calculated from 40 0.1-ha subplots in the primary forest of Tunas flat (TF), Tunas steep TS), Bonggo flat (BF), and Bonggo steep. 39

Table 3.2.   Mean basal area and standard error (in absolute and percentage terms) calculated from 20 0.1ha subplots in Tunas 4-year logged (T4), Tunas 8-year logged (T8), Bonggo 4-year logged (B4), and Bonggo 8-year logged stand. 39

Table 4.1.   Common tree families (DBH > 10 cm) with three and more species encountered in the primary moist forest of Tunas and Bonggo, Papua 42

Table 4.2.   The ten most important families (DBH>10cm) ranked based on Family Important Value (FIV) at two topographic stands in Tunas primary forest. 45

Table 4.3.   The ten most important families (DBH>10cm) ranked based on Family Important Value (FIV) at two topographic stands in Bonggo primary forest. 46

Table 4.4.   The ten most important tree species with DBH>10cm ranked based on their Important Value Index (IVI) of two different topographic conditions (flat and steep forest stands) at Tunas primary forest. 50

Table 4.5.   The ten most important tree species (IVI rank) with DBH>10cm found in two different topographic conditions (flat and steep forest stands) at Bonggo primary forest. 51

Table 4.6.   Diversity indices and their comparison between and within forests of Tunas and Bonggo in West Papua Indonesia 64

Table 4.7.   Species similarity between stands of Tunas and Bonggo forests presented in Jaccard-classic, Jaccard-individuals, and Jaccard-Basal Area calculated from tree species with DBH ≥ 10 cm. 66

Table 4.8.   Stem density, mean diameter, and BA of all trees with DBH ≥ 10 cm in four stands of Tunas and Bonggo forests. 71

Table 4.9.   The percentage and number of stems and species grouped
in five size classes of the four study stands 77

Table 4.10. Stand heights of all trees with DBH ≥ 10 cm in four stands of
Tunas and Bonggo forests 79

Table 4.11. Relative and absolute abundance of individuals and species
with DBH ≥10 cm distributed in three forest storeys in four
study stands of Tunas and Bonggo forests. 82

Table 4.12. The ten most important species of seedlings and saplings
ranked based on species' Important Value (IV) at two
topographic stands in Tunas primary forest. 90

Table 4.13. The ten most important species of seedlings and saplings
ranked based on species' Important Value (IV) at two
topographic stands in Bonggo primary forest 91

Table 4.14. Diversity measures of lower-canopy tree species of two
topographic stands in Tunas primary forest 94

Table 4.15. Diversity measures of lower-canopy tree species of two
topographic stands in Bonggo primary forest 94

Table 4.16. Chao-Jaccard abundance-based indices of similarity between
seedling, sapling, and canopy tree species in flat and steep
stands of Tunas primary forest. 96

Table 4.17. Chao-Jaccard abundance-based indices of similarity between
seedling, sapling, and canopy tree species in flat and steep
stands of Bonggo primary forest. 97

Table 4.18. The abundance of natural regeneration in flat and steep
stands of Tunas primary forest. 98

Table 4.19. The abundance of natural regeneration in flat and steep
stands of Bonggo primary forest 98

Table 4.20. Diameter and height of saplings at flat and steep stands in
Tunas and Bonggo forests 99

Table 5.1.   Ten most important canopy tree species (DBH > 10 cm) ranked based on the Importance Value Index (IVI), recorded in unlogged, 4-year and 8-year logged stands of Tunas forest 111

Table 5.2.   Ten most important canopy tree species (DBH > 10 cm) ranked based on the Importance Value Index (IVI), recorded in unlogged, 4-year and 8-year logged stands of Bonggo forest 115

Table 5.3.   Species richness and individual–species ratio of canopy trees at three stands (unlogged, 4-year logged and 8-year logged) in Tunas and Bonggo forests 119

Table 5.4.   Diversity indices of canopy trees at three stands (unlogged, 4-year logged and 8-year logged) in Tunas and Bonggo forests 121

Table 5.5.   The number of stems and species distributed in five size classes of the three timber commercial groups of Tunas stands 128

Table 5.6.   The number of stems and species distributed in five size classes of the three timber commercial groups of Bonggo stand 128

Table 5.7.   Ten most important regenerated species ranked based on Importance Value (IV), recorded in unlogged, 4-year and 8-year logged forest of Tunas 131

Table 5.8.   Ten most important regenerated species ranked based on the Importance Value (IV), recorded in unlogged, 4-year and 8-year logged forest of Bonggo 133

Table 6.1a.   Frequency of tree diameter classes and slope (b) of their linearised curves for every tree species (DBH >5cm) classified into five groups based on their recruitment and establishment ability in Tunas. 139

Table 6.1b.   The frequency of tree diameter classes and slope (b) of their linearised curves for every tree species (DBH >5cm) classified into five groups based on their recruitment and establishment ability in Bonggo 141

Table 6.2. $\delta^{15}N$ abundance (atom ‰) and total N (%) in the sapwood
tissue of various tree species classified into five groups
based on recruitment and establishment ability.............................149

Table 6.3.   Median test for $\delta^{15}N$ abundance (atom ‰) and total N (%) in
the sapwood tissue of five groups of tree species in the
Tunas and Bonggo forests. 152

Table 6.4. $\delta^{15}N$ abundance (atom ‰) and total N (%) in various parts of
trees from subtropical and tropical forests.....................................153

Table 6.5a. Light requirement characteristics of 50 tree species from
Tunas forest based on the slope of the diameter distribution
(b-value), depletion isotope $\delta^{15}N$ and total N...............................161

Table 6.5b. Light requirement characteristics of 53 tree species from
Bonggo forest based on the slope of the diameter distribution
(b-value), depletion isotope $\delta^{15}N$ and total N...............................165

List of Figures

Figure 2.1. Map of Indonesian Papua indicates two study sites, Tunas and Bonggo sites. 16

Figure 2.2. Climate diagram showing an average of 12 years (2001-2012) rainfall and temperature in the Tunas site. 18

Figure 2.3. Soil profile in Tunas forest 20

Figure 2.4. Climate diagram showing an average of 10 years (2002-2012) rainfall and temperature in the Bonggo site. 23

Figure 2.5. Soil profile in Bonggo forest 25

Figure 2.6. Arrangement of subplots. 28

Figure 2.7. Design of measuring plot. 29

Figure 2.8. Field work 34

Figure 3.1. Species accumulation curves for primary forest stands (A) and logged forest stands (B) of Tunas and Bonggo. The species name was recorded from all trees with DBH ≥ 10 cm. 37

Figure 4.1. Species rarefaction and extrapolation curves for four study stands of Tunas and Bonggo. 60

Figure 4.2. Observed and predicted stems in each 10-cm DBH class. 75

Figure 4.3. Diameter-height relationship of all stems with DBH≥10cm in Tunas flat (TF), Tunas steep (TS), Bonggo flat (BF), and Bonggo steep (BS) stands 86

Figure 4.4. Sapling distribution by diameter class at flat and steep stands in Tunas forest (A) and Bonggo forest (B) 101

Figure 4.5. Sapling distribution by height class at flat and steep stands in Tunas forest (A) and Bonggo forests (B) 103

Figure 5.1a. Tunas unlogged- (A) and logged-forest 4 years after logging (B) 107

Figure 5.1b. Tunas logged-forest 8 years after logging (C) 108

Figure 5.2.   Numbers of high-value, mixed and non-commercial tree
              species (DBH > 10 cm) recorded in unlogged, 4-year and 8-
              year logged stands in Tunas (A) and Bonggo (B) forests. 109

Figure 5.3.   Observed and predicted numbers of stems in 10-cm DBH
              size class intervals in unlogged, 4-year and 8-year logged
              stands in Tunas forest. 124

Figure 5.4.   Observed and predicted numbers of stems in 10-cm DBH
              size class intervals in unlogged, 4-year and 8-year logged
              stands in Bonggo forest. 125

Figure 6.1.   Correlation of $\delta^{15}N$ abundance (atom ‰) and total N (%) in
              the sapwood tissue of 50 tree species from Tunas forest (a)
              and 53 tree species from Bonggo forest (b). 151

Figure 6.2.   Correlation between the b-value and $\delta^{15}N$ atom (%) of 50
              tree species from Tunas forest (a) and 53 tree species from
              Bonggo forest (b). 157

Figure 6.3.   Correlation between the b-value and total nitrogen (%) of 50
              tree species from Tunas forest (a) and 53 tree species from
              Bonggo forest (b). 158

Abbreviations

^{15}N	:	Stable isotope nitrogen
A.s.l	:	Above sea level
BA	:	Basal area
BF	:	Bonggo flat stand
BS	:	Bonggo steep stand
CO_2	:	Carbon dioxide
DBH	:	Diameter at breast height (cm)
FAO	:	Food Agricultural Organization
Ha	:	Hectare
HPH	:	Hak Pengusahaan Hutan (Forest Concessionary Right) for timber logging in Indonesian natural forest
ITTO	:	The International Tropical Timber Organization
IUPHHK-HA	:	Ijin Usaha Pemanfaatan Hasil Hutan Kayu di Hutan Alam (Commercial Timber Forest Product Utilization Permit- In Natural Forest)
KZSI	:	Center for Stable Isotopic Research and Analysis University of Göttingen, Germany
m^2	:	Square meter
m^3	:	Cubic meter
Mha	:	Million hectare
MoFOR	:	Ministry of Forestry Republic of Indonesia
N	:	Nitrogen
N/ha	:	Number of trees per hectare
PPT	:	Pusat Penelitian Tanah (Indonesian Soil Research Center)
PT. TTL	:	Tunas Timber Lestari Ltd.
PT. WMT	:	Wapoga Mutiara Timber Ltd.
REDD	:	Reducing Emissions from Deforestation and Forest Degradation
RePPProt	:	Regional Physical Planning Programme for Transmigration, Ministry of Transmigration, Indonesia

RIL   : Reduced-impact logging
SD   : Standard deviation
SE   : Standard Error
SFM   : Sustainable Forest Management
SILIN   : Silviculture intensive
TF   : Tunas flat stand
TPI   : Tebang Pilih Indonesia (the Indonesian Selective Cutting
              System). The first Indonesian silviculture system for logging
              in concession areas in a natural mixed forest
TPTI   : Tebang Pilih Tanam Indonesia (the Indonesian Selective
              Cutting and Replanting System): the improved Indonesian
              silviculture system from TPI
TS   : Tunas steep stand
UNFCCC   : The United Nations Framework Convention on Climate
              Change
USDA   : United State Department of Agriculture

1. Introduction

1.1. The threat to tropical forests of insular South East Asia

Southeast (SE) Asia and the Pacific Archipelago (New Guinea and the Solomon Islands) harbour about 15% of the world's tropical forests (FAO, 1995). The region varies in forest type, including mixed deciduous forests, highly productive dipterocarp forests, and tropical moist forests (Richards, 1996; Thaman *et al.*, 2006). This tropical region is also home to carbon-rich ecosystems such as mangrove and peat swamp forests that are still located throughout coastal zones (Donato *et al.*, 2011; Page *et al.*, 2011). Moreover, the region's tropical forests play an essential role in environmental protection and biodiversity, socio-economy, and the living conditions of forest-dependent populations (Lee, 2009). Furthermore, these forests are of importance in the context of global carbon balance (Fox *et al.*, 2010; Page *et al.*, 2011).

A significant challenge to sustaining the tropical forests of SE Asia and the Pacific is rampant deforestation and forest degradation, driven by growing demands for food, timber, and other natural resources (DeFries *et al.*, 2010; Gibbs *et al.*, 2010; Foley *et al.*, 2011; Wilcove *et al.*, 2013). Forest cover in these regions decreased significantly between 1990 and 2010, with the vast majority of loss (about 2/3 of the loss experienced between 2000 and 2010) occurring in insular SE Asia, driven by logging and the use of peatlands for tree crops (in particular oil palm) (Houghton & Hackler, 1999; Miettinen *et al.*, 2011; Stibig *et al.*, 2014; Miettinen *et al.*, 2017). These pressures on forests are predicted to account for 13–85% of species losses in the region by 2100 (Sodhi *et al.*, 2010). Besides threatening existing endemic species in the region, reductions in forest cover will further accelerate CO_2 emissions from opened peatlands, impacting climate change (Miettinen *et al.*, 2011; Miettinen *et al.*, 2017).

The insular sub-region of Brunei, East Timor, Indonesia, Malaysia, the Philippines, Papua New Guinea (PNG) and the Solomon Islands, which together constitute approximately 70% of SE Asia's tropical forests, experienced rapid deforestation between 1990 and 2010 (Stibig *et al.*, 2014). The expansion of oil palm plantations during the 1990s and 2000s has been the greatest driver of deforestation in eastern Sumatra, coastal Sarawak, central and northeast Borneo, and southeast Papua (Miettinen *et al.*, 2017). There are also indications of the establishment of new forest plantations in Papua New Guinea, Peninsular Malaysia, Sumatra, and Sarawak (Stibig *et al.*, 2014). Indonesia has accounted for approximately 80% of forest conversion over the two decades (Stibig *et al.*, 2014). Sumatera and Borneo (Indonesia) accounted for approximately 10 Mha of the total 14.7 Mha forest cover losses in SE Asia between 2000 and 2010, with fibre (*Acacia mangium*) plantation and oil palm plantation concessions contributing the most (1.1 Mha and 1.2 Mha, respectively) (Abood *et al.*, 2015). Even though logging concessions have made a relatively small contribution to total deforestation, there might have been an occurrence of deforestation during logging operations (inside logging concessions) before their conversion to the tree and oil palm plantations (Casson, 2000; Kartodihardjo & Supriono, 2000). Before 2013, logging accounted for only marginal forest losses in Papua Indonesia, where deforestation was less than half that experienced by Sumatera and Borneo. Today, logging has become one of the leading drivers of industrial deforestation, and is responsible for increasing forest conversion on the island (Stibig *et al.*, 2014; Abood *et al.*, 2015; Austin *et al.*, 2017; Miettinen *et al.*, 2017).

Although the industrial sector might be expected to be the primary culprit, it was responsible for less than 50% (6.6 Mha) of Indonesia's deforestation (Stibig *et al.*, 2014). Some enterprise-based plantation activities have been shown to occur outside of concession boundaries (Gaveau & Salim, 2013). Nevertheless, small- and medium-scale illegal logging, land clearing, and forest fires, which also drive forest losses, should not be neglected (Curran *et al.*, 2004; Langner *et al.*, 2007; Ekadinata *et al.*, 2013). In some instances, protected forest areas are becoming isolated and even deforested, while buffer zones are getting degraded (Curran *et al.*, 2004) as the result of expanding populations and rapidly growing economies (Sodhi *et al.*, 2010). The factors described here are inter-correlated. Unmanaged (and/or illegal) logging methods may leave a forest in such poor condition that it becomes fragmented. Fragmentation effects and degradation caused by logging may leave residuals that are vulnerable to intentional and accidental fire and further encroachment, which in turn may lead to deforestation and conversion.

1.2. Regeneration of natural tropical primary forest

The main challenge of achieving sustainable forest management goals is the successful mimicry of pristine forest conditions through adjustment of harvesting procedures in conformance with tropical forests' natural regeneration processes. It is crucial to consider that virgin tropical forests vary in terms of tree species composition, diversity, and structure due to variations in the environmental setting. Topography and biogeography firmly control pollination and seed dispersal agents of tree species (Walker & Hope, 1982; Petocz, 1989; Sist *et al.*, 2003b). Differences in rainfall, humidity, and temperature (Richards, 1996) combine to contribute to the formation of varying soil types, soil textures, and nutrients, which are essential for the establishment and development of particular species (Gentry, 1988; Ashton & Hall, 1992; Tuomisto *et al.*, 2003). By comprehending the physical and biological factors that shape a forest community, one can more easily understand the historical, functional, and successional behaviours of forest ecosystems (Oliver & Larson, 1990; Spies, 1998) and better understand how to reproduce natural processes in an anthropogenically altered or created forest. This understanding can serve as an important reference in the process of planning forest management techniques such as harvesting and thinning (Spies, 1998). It can also inform optimal silvicultural practices in forest restoration and rehabilitation, as well as the identification of relevant data to support conservation programs (Teketay, 1995; Teketay & Bekele, 1995; Pyke *et al.*, 2001).

The tropical forest harbours hundreds of tree species. The tropics' tree communities are typically composed of a few species with many individuals (common species) and many species with one or two individuals (rare species). This composition allows for high species diversity in natural forests. Because many rare species not only occur locally (e.g., at the stand level) but also span an extensive biogeographical range, they may represent a vast absolute number of individuals at the landscape level (Primack & Hall, 1992; Poorter *et al.*, 1996; Pitman *et al.*, 1999). The mechanisms that allow many tree species to coexist in the same assemblage, in which the common species do not outcompete the many rare species, are vital to the natural regeneration of a tropical forest. In other words, the possibility of reproducing any individual tree depends on substantial factors of randomness in regeneration processes.

The number of tropical tree species is positively correlated with the number of pollinators and seed-dispersal animals: Insects, birds, bats, and various other mammals are vital to the successful reproduction of tropical trees (Howe & Miriti, 2000; Sist *et al.*, 2003b; Anitha *et al.*, 2010). Dispersion is essential to improving the probability of seed germination through the movement of seeds to more favourable sites that are far from the mother tree, such as tree-fall gaps or areas of nutrient-rich soil (Howe & Miriti, 2000; Babweteera & Brown, 2010).

The probability of tropical forests' successful regeneration mainly depends on small-scale natural disturbances that cause disordered canopy gaps (Denslow, 1987). Small openings in a canopy resulting from full or partial tree falls allow for the transmission of sunlight to seedlings and juvenile trees. Mid- and upper-canopy tree species mainly rely on this gap-light for their regeneration (Denslow, 1987; Kuusipalo *et al.*, 1997; Struhsaker, 1997). Tree species are grouped along a gradient of light dependence, from the light-demanding pioneers that depend on light throughout their lives to shade-tolerant types that may survive for many years in the full shade beneath the canopy (Sist *et al.*, 2003a; Vieira *et al.*, 2005; Herault *et al.*, 2010). Therefore, the species composition of forest recovering after disturbance varies with canopy-opening size and the species of seedlings or seeds that have been carried into the gap (Denslow, 1987; Howe & Miriti, 2000).

Primary tropical forest hardwoods, including most high-value timber species, are characterised by long lifespans and slow growth rates (Richards, 1996). These species mainly regenerate in a gap and adapt to shade over the course of their lives (Vieira *et al.*, 2005). Seedlings of the commercially high-value mahogany require sunlight during their early years but can tolerate shade after becoming established (Sist *et al.*, 2003b; Hall, 2008). In fact, many dipterocarps of Southeast Asia tolerate shade throughout most of their lives (Kuusipalo *et al.*, 1997; Sist *et al.*, 2003b). Tropical timber trees are often hundreds of years old at harvest; some species with high wood density reach as much as 1000 years (Kurokawa *et al.*, 2003). Tree species vary markedly in diameter increment between and within tree species, depending on various factors such as age, edaphic and climatic conditions, and associated lianas (Peña-Claros *et al.*, 2008; Toledo *et al.*, 2011). Although most tropical timber species have innately low growth rates (Dauber *et al.*, 2005; Vieira *et al.*, 2005; Valle *et al.*, 2007), they are capable of responding swiftly to light provided by canopy gaps and increasing their diameter increment (Herault *et al.*, 2010).

The huge number of tropical tree species is reflected in the population structure of the forest assemblage. The long tail part of the inverted J-shape curve of size class distribution represents larger stems with few seedlings, saplings, and poles. For instance, 46 out of 49 valuable timber species from the Bolivian forest are large-sized trees with low numbers of individuals and low regeneration rates; these are mostly pan-neotropical species, such as *Swietenia macrophylla* and *Cedrela fissilis* (Fredericksen, 1999). Valuable timber species with low regeneration rates are common throughout the Amazon, as well as in African and Southeast Asian forests (Poorter *et al.*, 1996; Kammesheidt *et al.*, 2001; Sist *et al.*, 2003b; Hall, 2008; Schulze *et al.*, 2008; Anitha *et al.*, 2010; Babweteera & Brown, 2010). Their low population density is mainly associated with their status as nonpioneer light-demanding species whose seedlings cannot tolerate shade, causing high seed and seedling mortality rates in natural forests (Hall, 2008). The sparse regeneration implies a scant probability of tree recruitment in a high-diversity assemblage. However, their dispersal ability allows them to scatter across the natural forest such that at a landscape level, they can persist as a viable population.

1.3. Logging impacts on tropical forests

Forest destruction caused by logging in tropical forests varies depending on the harvest intensity level, indicated by the number or volume of timber harvested per ha. When logs are felled and dragged ('skidded') from their stump to the log landing, they create tree gaps and skid trails. Tree felling can destroy surrounding trees through collateral canopy damage and skidding processes, which clear seedlings and saplings and injures future crop trees, often leading to mortality (Hawthorne *et al.*, 2011). Thus, the magnitude of forest damage increases with the number of felled trees to a point at which where primary forest, non-pioneer trees, cannot naturally grow (Struhsaker, 1997; Sist & Nguyen-Thé, 2002; van Gardingen *et al.*, 2006). Under high logging intensity conditions, in which more than eight trees are typically extracted per ha (Sist *et al.*, 2003a), the pressure on the natural regeneration of primary forest substantially kills most of the future crops and seed trees (Verissimo *et al.*, 1995; Sist *et al.*, 1998). Felling creates gaps that strongly favour fast-growing pioneer vegetation that easily outcompetes slow-growth species (Kuusipalo *et al.*, 1997; Struhsaker, 1997; Sist & Nguyen-Thé, 2002; Sist *et al.*, 2003b; Park *et al.*, 2005). This may also have a detrimental influence on populations of animal pollinators and seed dispersers (Struhsaker, 1997; Babweteera & Brown, 2010).

Most of the tropical commercial timber species are non-pioneer, light-demanding species. During the recruitment stage (up to a DBH of 10 cm), trees' growth rates are optimal when canopy openings are few and relatively little soil disruption occurs (Kuusipalo *et al.*, 1997; Hall, 2008). Therefore, the gap sizes created during felling and skidding are vital to a balance between sustaining valuable tree species and restricting pioneer vegetation growth in logged forests. In a less than 500 m^2 felling area, the promotion of non-pioneer, shade-intolerant juveniles were found to be ideal, with a maximum of 10% stand canopy opening and a minimum of 85% maintained stand basal area (Struhsaker, 1997; Sist *et al.*, 2003a). In other words, the growth of valuable timber species is optimized by low-intensity logging. The practice of cutting less than five stems per ha has been suggested to minimize adverse impacts on future trees and drive stand recovery toward initial composition and structure (Struhsaker, 1997; Sist & Nguyen-Thé, 2002; ter Steege *et al.*, 2002; Hall *et al.*, 2003).

However, the current Indonesian silvicultural system has less control over cutting-intensity if an intact forest with large high-value timbers is the targeted logging area (Sist *et al.*, 1998). The Bornean dipterocarp forest, where commercially valuable trees are in abundance (about 200 species), is mainly harvested at high intensities of more than ten trees per ha, which results in 50% canopy gaps and severe damage on the residual stand (Sist & Nguyen-Thé, 2002; Sist *et al.*, 2003a; Forshed *et al.*, 2008). Most of these logged forests remain low in timber value; many are eventually termed "degraded forests" and reclassified for other land uses (Kartodihardjo & Supriono, 2000; Gaveau *et al.*, 2013).

Even though patchily distributed and represented by only three genera (*Anisoptera, Vatica,* and *Hopea*) and about five species, the Dipterocarpaceae family dominates the forest canopy in South-West New Guinea (Paijmans, 1976). In these forests, approximately ten out of every fifteen commercial trees cut per ha are dipterocarps, which has led to the domination of 8-year logged stands by the pioneer, light-demanding family *Moraceae*, followed by the dipterocarp *Vatica rasak* (Gandhi & Mitlöhner, 2014). Similarly, the heavily exploited Merbau tree (*Intsia bijuga* O.K.) in the northern part of the island (Thaman *et al.*, 2006; Sadono, 2014) has been harvested at high intensities, constituting approximately five of every thirteen harvestable trees (personal interview). This high-value species has also attracted destructive, illegal logging in the Papuan forest (Newman, 2008). Currently, the disappearing Merbau stock is triggering a demand to classify the species as endangered in APPENDIX III CITES. Should the current Indonesian logging practices, which fail to meet the sustainability qualifications, continue, subsequent harvests of less-valuable species within these forests raise the threat of reclassification to "degraded forest", as has occurred in other parts of Indonesia. In short, a trade-off between the number of each tree species cut above which the species locally vanishes is urgently necessary.

1.4. Sustaining tropical logged forests

Sustainable Forest Management (SFM) has been broadly defined. It ranges from the simple-narrowed that is 'management capable of maintaining timber productivity in a management cycle' to the modern-broadly as 'management preserving productivity, forest structure, diversity, and the fundamental ecological process of populations, communities, and the ecosystem' (FAO, 2014). SFM has been of global relevance in the effort to combat large-scale, industrially driven deforestation in the tropics, and SFM is one of the United Nations' instruments for addressing sustainable development and biodiversity conservation. The UN Framework Convention on Climate Change (UNFCCC) provides incentives for the implementation of SFM in tropical forests through its Reducing Emissions from Deforestation and Forest Degradation (REDD+) policy (UNFCCC, 2008). Several tropical countries, mainly members of the International Tropical Timber Organization (ITTO), have demonstrated a serious commitment to SFM as a means of preserving tropical forests (Blaser *et al.*, 2011). For example, Indonesia has established its Criteria and Indicators, developed national standards for forest certification, and introduced the mandatory verification of SFM into its management system (ITTO, 2006). The country's revised Criteria and Indicators for SFM consist of four criteria (enabling conditions, production, ecology, and social aspects) and 24 indicators, endorsed by the national government through Ministerial Forestry Regulation No. 4795/Kpts-II/2002 issued on 3 June 2002.

It is debatable whether the practice of harvesting high-value timber for more than a single cutting cycle can assure the sustainability of the natural forest ecosystem, as stated by SFM goals (Nasi & Frost, 2009; Zimmerman & Kormos, 2012). As previously discussed, many tropical trees of intact forests, including high-value timber species, have life spans and growth rates such that they take years to become harvestable. These species' seeds and seedlings also experience high mortality rates. At last, these timber species rely on a diversity of animal dispersers for reproduction. All of these traits imply that vast, continuous areas of forests are important in maintaining viable population sizes and enabling sustainable timber harvesting in tropical forests (Pitman *et al.*, 1999). In general, however, most of the timber concession areas are limited to one cycle period of 30-35 years, which compels timber companies to reenter the same places for subsequent cutting cycles (Zimmerman & Kormos, 2012). It is a considerable challenge for profit-oriented companies with limited knowledge of the tropical trees' life history to maintain the productivity of high-value timber using current cycles (Zimmerman & Kormos, 2012). Therefore, without implementing financial incentives and new forest regulations, sustainability goals are unlikely to be achievable (Schulze, 2008). The problem is exacerbated if little is known about the interactions of individual trees with their environment, e.g., species-specific growth rate, germination and growth requirements, animal pollinator and disperser species, and edaphic preferences (de Freitas & Pinard, 2008; Grogan *et al.*, 2008).

Although most countries in the tropics have a standard protocol for timber logging that includes anticipation of drivers and impacts (ITTO, 2006), the life histories of timber species are universally poorly understood (Zimmerman & Kormos, 2012). The high international demand for tropical timber has led to increasingly profit-oriented motivations and the harvesting of valuable species first. When these species are less available for exploitation in the second cycle, the next lower-grade timbers are targeted, until the commercial species are threatened by disappearance altogether, and the land becomes more financially viable for uses other than forestry (Casson, 2000; Asner *et al.*, 2006). The voracious consumption of commercial trees in the export market has forced the expansion of logging areas far into the pristine forests of the Amazon, Central Africa, and Borneo (Kammesheidt *et al.*, 2001; Laporte *et al.*, 2007; Hall, 2008; Bryan *et al.*, 2010). To compound the problem in some instances, abundant timber supply from illegal and conventionally unmanaged logging diminishes the incentive for management or conservation (Nawir & Rumboko, 2007). Conventional logging has reduced many Southeast Asian primary forestlands to degraded secondary forests, some of which have been cleared for agriculture (Casson, 2000; Kartodihardjo & Supriono, 2000; Stibig *et al.*, 2014; Abood *et al.*, 2015). This relentless process is also well underway in the wild forests of the last frontiers of Indonesia, including the Papuan forest (Austin *et al.*, 2017; Austin *et al.*, 2019).

SFM protocols that rely on government-mandated silviculture systems have been practised in many tropical timber-producing countries. Blaser *et al.* (2011) reported that many ITTO countries practice cutting cycles (30-35 years), minimum felling diameters (40-60 cm DBH for all species), per-unit-area harvest intensities, and seed tree retention rates applied in combination with proven techniques for reducing damage to the residual stand during logging operations (reduced-impact logging, RIL). The report also indicates that silvicultural techniques that can enhance the abundance and diameter increment of timber species' seedlings and saplings for recruitment are not yet clearly stated in SFM procedures.

RIL is a set of timber-harvesting techniques for reducing the mortality rates of residual trees caused by logging and preserving the ecological integrity of the logged stand. These techniques are useful for increasing the survival of seedlings, saplings, and future trees of commercial species, and have proven effective in reducing collateral damage to the residual stand by 20%–50% (Putz *et al.*, 2008). However, if logging at higher intensities (>five trees ha^{-1}) and employing diameter-limit cutting RIL alone fail to sustain production and take no marked effect in future harvests (Kammesheidt *et al.*, 2001; Sist *et al.*, 2003b; van Gardingen *et al.*, 2006; Peña-Claros *et al.*, 2008; Kukkonen & Hohnwald, 2009).

Furthermore, some studies have indicated that logging operated under current SFM protocols is inconsistent with sustainability in terms of production and ecological integrity (Poorter *et al.*, 1996; Sist *et al.*, 2003b; Dauber *et al.*, 2005; de Freitas & Pinard, 2008; Schulze *et al.*, 2008; Herault *et al.*, 2010). These researchers show that minimum rotation cycles (typically 25-35 years) are too short to maintain the quality and quantity of the first yield in the third cycle, minimum felling diameters (50 cm DBH) are too small to preserve adequate populations of reproductive adults and control harvest intensity, harvest intensities (around eight trees per ha) are too high and leave residual stands with overgrowths of pioneer vegetation that outcompete crop trees, and seed-tree retention rates (10% retention of trees greater than 45 cm DBH) are inadequate. Together, these studies suggest that minimum felling diameter and commercial retention regulations are not appropriate for all timber species' life histories. Instead, a species-specific method that incorporates local intact forest structure, growth rates, and relative population densities is in order. Unfortunately, little information is available for many forests on species-specific life history, composition, and structures.

1.5. The rationale of the study

The expansion of forest conversion in the insular regions of tropical Southeast Asia is of real concern. Authorities may promote the conservation of pristine forests strictly for the purpose of preserving biological diversity or to sustain other values and functions (Chape *et al.*, 2005; Sodhi *et al.*, 2010). Although restrictions on timber extraction and agricultural practices are high in conserved areas, socio-economic factors and the cost of protection are intricately related in these forests, often resulting in gradual encroachments and the creation of ecologically degraded parcel habitats (Joppa *et al.*, 2009; Wilson *et al.*, 2010). These threats have led to increasing attention being given to opportunities for conservation in non-protected forests. Promoting conservation initiatives in natural timber concessions and combining these with already existing programs in protected forests has been suggested as a viable means of providing a more spacious forest landscape ecosystem and slowing the rate of forest conversion (Clark *et al.*, 2009; Wilson *et al.*, 2010; Edwards & Laurance, 2013).

The underlying economic incentive behind conservation in timber concession areas is the maintenance of sufficient forest cover to ensure continued profitable logging and prevent the conversion of forests into cropland. Gaveau *et al.* (2013) assert that logging concessions in the rapidly deforested island of Borneo are capable of demonstrating similar competence as protected areas in maintaining forest cover. Fisher *et al.* (2011) outline a highly cost-efficient way of preserving high levels of biodiversity in selectively logged forests as a means of enlarging existing protected areas. Although the timber standing stocks decline by five times after two cutting cycles, the new joining shows vast promise in protecting forest landscapes and providing continued timber production. Opponents of this scheme have claimed that forest protection within state-designated concessions is impractical due to inadequate enforcement (Indarto *et al.*, 2015) and facilitation of illegal encroachment and hunting in areas where road networks have been left by logging practices (Laurance *et al.*, 2002; Pfaff *et al.*, 2007; Gaveau *et al.*, 2009; Zimmerman & Kormos, 2012). Other pessimistic assessments argue that logging concessions will provide protection only for a relatively short period and that as timber stocks decline and demand for agricultural land increases, financial necessity will ultimately push authorities to reclassify logged forests to crop plantations (Kartodihardjo & Supriono, 2000; Zimmerman & Kormos, 2012).

A better proposition for natural timber concessions as protected areas is to improve timber yields while retaining ecological integrity (conservation values) inside the forest. A recent global meta-analysis of more than 100 scientific papers reveals that tropical logging leaves about half of the initial timber stocks, three-fourths of the carbon stocks, and 85–100% of mammal, bird, invertebrate, and plant species richness after the first logging cycle (Putz *et al.*, 2012). If efforts to reduce collateral damage and implement post-harvest treatments are intensive, potential productivities can be successfully augmented and sustained (Putz *et al.*, 2012). Others suggest intensive silviculture techniques, i.e., indirectly lowering high cutting intensity, ensuring a single tree gap, increasing minimum diameter cutting limit, and improving growth rate (through liana clearing and girdling of competing trees) (Sist *et al.*, 2003a; Villegas *et al.*, 2009). However, if the SFM does not include the silviculture techniques in the protocol, including longer rotation cycles for large-scale logging industries, sustainable yields and richness of high-value timber species will not occur after third-harvest (Zimmerman & Kormos, 2012). Instead, knowledge handover to small-scale forest resource managers by forestry-trained locals has proven to be the most promising means of securing the world's remaining tropical forests (Bray *et al.*, 2003).

The present study focuses on logging concessions in the natural forests of Papua, an Indonesian territory in the western part of New Guinea. Papua covers approximately 41.51 Mha, with forest comprising most (71.64%) of the area (Stibig *et al.*, 2014; Abood *et al.*, 2015). New Guinea is an outstandingly biodiverse region (Petocz, 1989; Beehler, 2007; Polhemus & Allen, 2007; Takeuchi, 2007). The mixing of original Gondwanan with subsequent Laurasian flora after the collision of the Australian and Southeast Asian plates about 10 million years ago could explain this high level of diversity (Walker & Hope, 1982). Since then, the island's continual uplift has led to environmental instability and physiographic features that have profoundly affected the overall distribution and isolation of the modern island's flora and fauna, which is characterised by rapid speciation (Walker & Hope, 1982). Based on the distribution of mountains and lowlands, which have acted as barriers to the dispersal of some organisms and as sites for speciation and diversification, Beehler (2007) defined 13 bird region-based biogeographic subunits for species endemism areas in Papua. These subunits overlie those delineated by Polhemus & Allen (2007) for 20 freshwater biogeographic subunits in most places. The protected area system's design for conservation and development of Papuan forest utilization adopted until recently has also considered the island's physiographic features and species endemism (Petocz, 1989).

The northwestern and southwestern lowlands of New Guinea are two extensive biogeographic subregions characterised by low species endemism because of their weak physical isolation and wide-range of species (Beehler, 2007; Polhemus & Allen, 2007). Nonetheless, the two lowlands' forests host various high-valued timber species (Paijmans, 1976; Petocz, 1989). These forests, in which trees may exceed 50 m in height, are mainly composed of high-value timber species such as *Alstonia, Intsia, Pometia, Terminalia,* and some *Dipterocarps* (i.e., *Anisoptera, Vatica, and Hopea*) (Paijmans, 1976). Accordingly, the forestry ministry of Indonesia, through Act No. 782/Menhut-II/2012 and 783/Menhut-II/2014, has allocated about 94% (20.26 Mha) of the total Papuan lowlands as permanent production forest (14.67 Mha) and convertible production forest (5.59 Mha). Both of these classification systems refer to state-owned forests designated for production purposes, but only convertible production classification allows for conversion to other forest land uses. Since 1970, the government has established 60 timber concession units across approximately 10.75 Mha (73%) of Papuan production forests; the number of concessions decreased to 51 units, but the overall area increased to 11.58 Mha (79%) in 1997 (MoFOR, 2008). Currently, 38 logging concession units covering about 5.60 Mha are operative; another 6.0 Mha are classified as "nonactive" concessions (MoEFOR, 2017). The authorities have converted some of these nonactive-areas (83.1 Kha) to non-forestry land uses such as transmigration, tree, and tree crop plantations during the 2000s (BPKH Wilayah X, 2009). Apart from this "small" forest conversion, some studies report substantial increases in Papuan intact forest losses between 1995 and 2015; even though relatively small in terms of overall forest loss in Indonesia, all together they are projected to become the dominants means of industrially driven deforestation in Papua in the future.

The prospect of future loss of the highly productive Papuan lowland forests is troubling. The New Guinean lowlands are estimated to be home to about 1,200 tree species from around 80 families, among an estimated number of 20,000 Papuasian plant species (Womersley *et al.*, 1978), as well as the near-endemic bird of paradise (*Paradisaea apoda*), the mammal (*Uromys scaphax*), the affiliated-endemic fig-parrot (*Psittaculirostris salvadorii*) and friarbird (*Philemon brassii*), and various non-documented mammals (Beehler, 2007), as well as other essential endemic faunas (Allison, 2007). Importantly, timber concessionaires of designated production forests are strategically positioned to slow and protect again activities that potentially encroach upon neighbouring protected forests. For example, the Nature Reserve of Goutier-Foja Mountain in the North Coastal Ranges is directly adjacent to the northern lowland's production forests, which help regulate accessibility to the nature reserve. It is essential to preserve the conservation values of Papuan selectively logged forests through the SFM mechanisms to preserve the forests as forests, and protect them from conversion to other uses.

SFM calls upon a forest management unit (e.g., logging concession) to maintain initial forest condition parameters (e.g., tree species composition, diversity, and stand structure). The attempt to promote sustainability of logging activities in the Papuan forest is in its infancy, having begun only recently through the implementation of RIL and Silviculture Intensive (SILIN) programs in the Indonesian silviculture system (TPTI). Thus, very little data exist on the life histories and local population structures of timber species. Therefore, the present study examined the characteristics of primary forest in two Papuan logging concessions, including tree species composition and diversity, stand structure, regeneration, and species light requirement. The study also assessed the logging impacts on stand characteristics in the two timber concession areas. Based on the differences in stand characteristics between intact and logged forests, the study evaluated and recommended tree species performance for the silviculture approach in TPTI.

The objectives of the study were, firstly, to analyse forest stands' initial condition, (including tree species composition, diversity, regeneration, and stand structure in two logging concessions), secondly, to determine species-specific dynamics and light requirements, and thirdly, to evaluate differences between intact- and logged-stand characteristics. Finally, the study provided essential information for SFM-compliant silvicultural practices vital to retaining the productive potential of Papuan selectively logged forests.

2. Study sites and Methods

2.1. Site selection

In order to be able to access intact forests of interest in remote areas and assess their responses to logging practices, a timber concession area was selected as the study site. In addition to describing characteristics of the natural primary and disturbed lowland forests in different evolution of tectonic provinces, the study forests must be a representation of the Australian Stable Province in the south and Mobile Belt Province in the north. Accordingly, data collection in this research was assigned to two different sites: Tunas forest, representing the southern lowland of the Stable Province, and Bonggo forest, characterizing the northern coastal lowland of the Mobile Province (Figure 2.1).

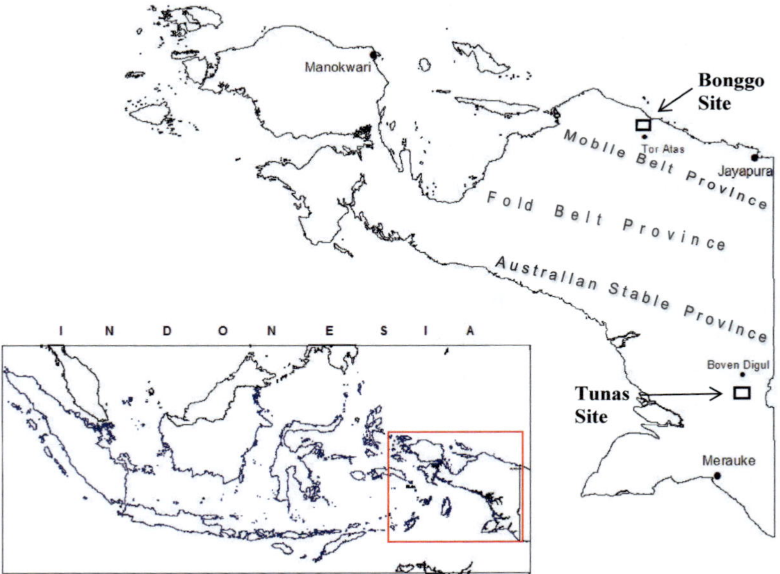

Figure 2.1. Map of Indonesian Papua indicates two study sites, Tunas and Bonggo sites.

2.2. Tunas site

2.2.1. Location

The southern study site in the Tunas forest was situated inside the logging concession area of PT. Tunas Timber Lestari (PT. TTL). The Forest Management Unit of PT. TTL administratively covers part of the territories of Jair and Mindiptanah Districts in the Boven Digoel Region, Papua Province, Indonesia. The confluence of the Uwim-Merah, Muyu, and Fly rivers marks the natural border of the logging concession in the south. It extends northward along the Muyu river in the west and the national border of Indonesia-Papua New Guinea in the east up to the concession border of PT. Citra Lembah Kencana II. The area is located between $140^0\,21'\,00"$ - $140^0\,59'\,00"$ E and $05^0\,50'\,50"$ - $6^0\,42'\,00"$S.

2.2.2. Climate

Twelve years of rainfall and temperature recorded by the meteorology and geophysics station at Tanah Merah, Boven Digul (approximately 73 km from the study sites) were analyzed and presented in a monthly climate diagram (Walter & Lieth, 1967), as depicted in Figure 2.2.

According to the classification of tropical lowland climatic type (Richards, 1996), the Tunas site belongs to the tropical wet class, as indicated by the humidity index of 19. It has high annual rainfall, exceeding 3000 mm, with two distinct seasonal monthly rainfalls. The very wet months (>200 mm) occur between October and April, with the peak monthly precipitation of 455 mm in March, and the wet months (100-199 mm) from May to September. The long-wet session of 5 continuous months was the rationale for placing this study site in a tropical moist forest group. The temperature at the Tunas site was constant throughout the year, with a mean annual temperature of 27.3^0C and a yearly variation of less than 2^0C.

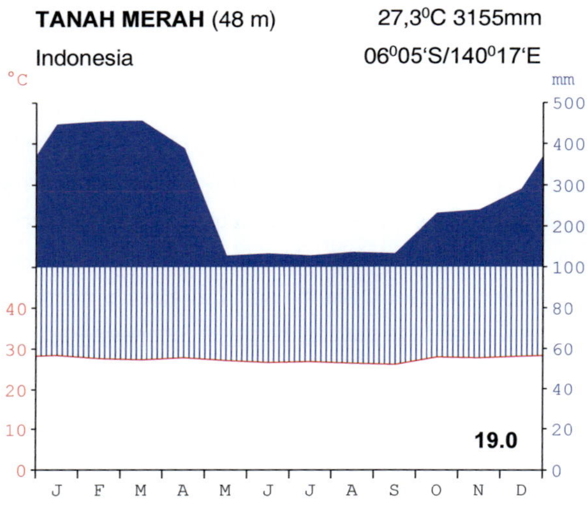

The data is a collection of Meteorology and Geophysics station at Tanah Merah, Boven Digoel. The diagram is developed using the ecological climate diagram by Walter and Lieth (1960).

Figure 2.2. Climate diagram showing an average of 12 years (2001-2012) rainfall and temperature in the Tunas site.

2.2.3. Topography and hydrology

Situated about 80 km southward from the foothills of the central mountain range to the southern plain developed from the Australian Continental plate, the study area gradually declines in altitude from 200 to 30 m asl. to form a gradual slope, from gently steep (8-15%) in the north to relatively flat (2-8%) in the south (RKU PT. TTL, 2009). Flanked by the Uwim Merah and Fly rivers, the southwest portion of the relatively flat area is temporarily inundated when rainfall is high. It covers just above 15% of the total area. The rest area is unflooded throughout the year, with more creeks flowing from the highest place between the two main rivers of Muyu and Fly.

2.2.4. Geology and soils

Soils in the Tunas site have developed from two geological formations. One is an old riverine sedimentary formation that covers most of the area. The other is a young riverine sedimentary formation located at the bench of Fly and Uwim Merah rivers (RePPProT, 1987).

In general, five factors influence soil formation: climate, vegetation, relief, parent material, and time. Variation in these factors contributes to differences in soil formation, even over relatively close distances. There are two soil groups at the study site, based on the Indonesian soil map (Puslitbangtanak, 2000) and the South East Asia standard map from the Food and Agriculture Organization (FAO-Unesco, 1974). The first is reddish yellow podzolic (classified by Puslitbangtanak, 2000) or ferric acrisols (classified by FAO-Unesco, 1974) or ultisol (USDA soil taxonomy, 1998), which is strongly weathered, acid and of low fertility. This soil type is mainly found in the Tunas forest; the profile is shown in Figure 2.3. The second is alluvial (Puslitbangtanak, 2000) or eutric fluvisols (FAO-Unesco, 1974) or entisol (USDA soil taxonomy, 1998), which are very young soils with little or no profile and mainly occur on recent alluvium or on steep slopes where soil erosion takes place (Hope & Hartemink, 2007). This soil increases in acidity when far from the coastal line.

Figure 2.3. Soil profile in Tunas forest

2.2.5. Forest history and resources

Hak Pengusahaan Hutan/HPH (forest concessionary right) of Tunas site, covering an area of about 200.000 ha, was issued to PT. Tunas Sawaerma in 1993. Conventional timber harvesting began with a limited TPI system applied due to the imaginary HPH border leading to the incertitude of the work plan map and less supervising control from the government. After 15 years, the ground-marked borderline of the HPH was completed, and all forest management block-maps were successfully updated using satellite imagery. During that period, the management plan was gradually improved, allowing the concessionary in 2009 to an extension of its *Ijin Usaha Pemanfaatan Hasil Hutan Kayu di Hutan*

Alam/IUPHHK-HA (Commercial Timber Forest Product Utilization Permit - In Natural Forest). The IUPHHK-HA covered approximately an area of 214,935 ha through the issuance of the Indonesian Ministry of forestry Act No. SK.101/Menhut-II/2009. Due to management reasons, the following year, the company's name was changed to PT. Tunas Timber Lestari (PT. TTL) under an act of the Indonesian Ministry of Forestry No. 711/Menhut-II/2010.

RKU PT. TTL (2009) reported that the company logged an area of 115,271 ha, or 53.6% of the total concession area, in the years from 1993 to 2009, and 41.774 ha (19.4%) of primary forest persisted for a subsequent cutting period. According to its IUPHHK-HA, PT.TTL has been allowed to harvest the primary forest at a rate of 2,408 ha year^{-1} (90,194 m^3 year^{-1}). This means that by 2026, the primary forest at the Tunas site is predicted to be converted to a Logged over Area (LOA). Apart from the timber forests, there are non-forest areas (areas containing fewer woody trees and shrubs, mostly found in swampy areas) covering about 44,606 ha (20.8%), and bodies of water covering 11,697 ha (5.5%).

Tunas Forest is dominated by valuable commercial species such as *Vatica rasak* and *Canarium asperum*; other commercial species include *Syzigium* sp, *Litsea timoriana,* and *Planchonella urophylla*, and non-commercial species include *Actinodapne nitida, Beismedia* sp, *Blumeodendron amboinicum, Linociera macrophylla, Meduchantera* sp, and *Pimeliodendron amboinicum* (Kuswandi, 2010). It has been reported that the potential tree density or volume of valuable commercial species ready to be harvested (DBH > 40 cm) was 9.69 stems ha^{-1} (32.99 m^3). That of future trees (DBH 20-40 cm) was 32.20 stems ha^{-1} or 20.96 m^3 ha^{-1}. The potentials of commercial species were found to be less than those of valuable commercial species, which were 6.13 stems ha^{-1} (17.83 m^3 ha^{-1}) and 12.93 stems ha^{-1} (11.22 m^3 ha^{-1}), respectively, for harvestable and future sized trees (PT. TTL, 2011).

2.3. Bonggo site

2.3.1. Location

The northern study site was located in the concession area of PT. Wapoga Mutiara Timber Unit II (PT. WMT-II). The logging area of PT. WMT-II belongs to the Bonggo and Pantai Timur districts of the Sarmi Region, Papua Province, Indonesia. The IUPHHK-HA area borders the Pacific Ocean in the north, IUPHHK-HA PT. Kebun Sari Putra and PT. Salaki Mandiri Sejahtera in the south, Toarim River in the east, and the Biri River in the west. The area is located between $139^0\,08'\,00''$ - $139^0\,48'\,00''$ E and $02^0\,08'\,00''$ - $02^0\,38'\,00''$ S.

2.3.2. Climate

Rainfall and temperature data over ten years (2003-2012) from the meteorology and geophysics station in Sarmi City (approximately 85 km from the study sites) were analyzed and presented in a monthly climate diagram (Walter and Lieth, 1960), as can be seen in Figure 2.4.

According to the tropical lowland climatic type (Richards, 1996), the Bonggo site is classified in the tropical, very wet class based on the humidity index of 21. It has a high annual rainfall of 3412 mm, and a short-wet period (100-99 mm) between October and December. Very wet months (>200 mm) occurred over eight months from January to September, with a maximum of 484 mm in July. This long, very wet season classifies the Bonggo site as a tropical, very moist forest group.

The temperature in the Bonggo site was reasonably constant throughout the year, with a mean annual temperature of 27.8^0C and a monthly variation of not more than 2^0C.

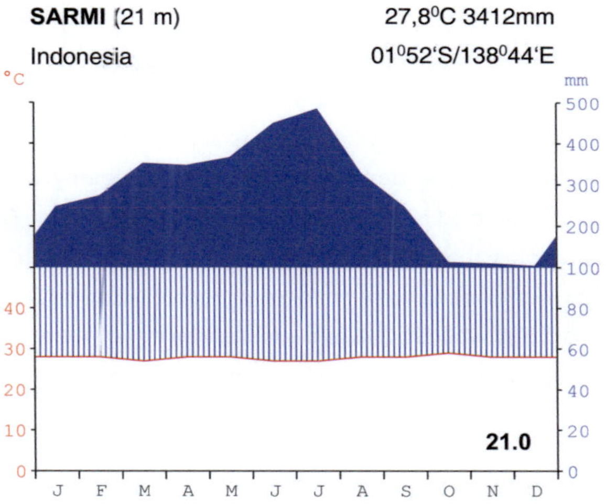

The data is a collection of Meteorology and Geophysics station at Sarmi, Papua. The diagram is developed using the ecological climate diagram by Walter and Lieth (1960).

Figure 2.4. Climate diagram showing an average of 10 years (2002-2012) rainfall and temperature in the Bonggo site.

2.3.3. Topography and hydrology

The Bonggo site's topography varies from relatively flat to undulating at a range of altitudes from 50 to 300 m asl. 63.81% of the area occupies hilly undulating land with steep slopes of 26-40%. The rest of the area lies in the coastal lowland, comprised of relatively flat land (18.44%, with slopes of 2 - 8%) and gently dipping land (17.75%, with slope range between 9 and 15%) (RKU PT. WMT, 2012).

Important rivers such as Tor, Bier, Kwaritor, Biri, Wiru, and Toarim drain the concessionary of PT WMT-II. Some small streams also flow into these main rivers. Certain valleys, where the streams are slowed and trapped, are temporarily inundated when heavy rain occurs close to the coastal lowland.

2.3.4. Geology and soils

The Bonggo study site is located in the Mobile Belt Tectonic Province, where the northern lowland range and coastal lowland were emplaced during the Oligocene and Pliocene, as mentioned elsewhere. This tectonic province consists of mixed arc terranes that were laminated to Northern New Guinea by a series of arc collisions and incorporated by arcs created from the margin of the Caroline and Solomon Sea plates that moved far from the East (Polhemus, 2007). For instance, a block of Miocene volcanic rocks is present along the Southern coast of Cenderawasih Bay. Various lithological formations reflect the higher degree of complexity in the arc amalgamation of the Mobile Belt, as compared with the central range Fold Belt and Australian Stable Platform provinces.

Arc collisions with continents or other arcs are typically characterized by the emplacement of a distinct stratigraphic assemblage. Polhemus (2007) indicated stratigraphically intact accreted arc terranes of Northern New Guinea often exhibit sequential bands of tectonic melange (derived from the fore-arc ridge), limestone (derived from the floor of the fore-arc basin), basalts, and other volcanic materials (derived from volcanic arcs), and ophiolite (derived from the back-arc basin). Geology maps of RePPProT (1987) and Papuan geology maps compiled by Petocz (1989) indicate that the Mobile Belt is comprised of sedimentary rocks (tectonic melange and limestone) and igneous rocks (volcanic and plutonic rocks).

Therefore, these parent materials (sedimentary and igneous rocks) play an essential role in soil formation at the Bonggo site, in combination with weathering, climate, vegetation, topography, and time.

As a result, the following soil orders based on USDA soil taxonomy (FAO term given in parentheses) occur at the Bonggo site (FAO-Unesco, 1974; RKU PT. WMT, 2012), with more important groups indicated in italics (RePPProT, 1987).

Ultisols (acrisols) are strongly weathered and acidic soils. *Tropudalfs* are the more important group found on gentle slopes (humic acrisols) and valley floors with meandering rivers (ferric acrisols). It has a yellow mottled horizon grading to clay and is of low fertility.

Inceptisols (district and humic cambisols) are moderately weathered soils with a slightly developed horizon. *Eutropepts* have rich organic matter, develop on mud volcanic rocks, and are located on gently steep slopes, while *dystropepts* are moderately high organic-matter soil developed on plutonic rocks on steep slopes.

Figure 2.5. Soil profile in Bonggo forest

2.3.5. Forest history and resources

Logging activity of PT. Wapoga Mutiara Timber (PT. WMT) has occurred since 1990, based on acts No. 744/Kpts-II/1990, dated 13th December 1990. The area stated on the acts covered an area of 357,700 ha and was separated into two different locations. PT. WMT Unit-I occupied an area of 178,800 ha between Kuri River and Umar Bay, Manokwari Region (recently Teluk Wondama Region of West Papua Province) during PT. WMT Unit-II is situated in an area of 196,900 ha between the Toarim and Biri rivers, Jayapura Region (recently Sarmi Region of Papua Province). After the first 20-year period, PT. WMT was given a 45-year concessionary extension starting in 2011, based on the Indonesian Ministry of Forestry Act No. 723/Menhut-II/2011 dated 20th December 2011. The concession area allowed was less than that in the previous period. PT. WMT Unit-I and Unit-II manage 130,715 ha and 169,170 ha, respectively. As there was difficulty in managing two distant areas in one management, PT. WMT proposed an addendum to split its concessionaries into two different Forest Management Units. By 7th November 2012, IUPHHK-HA PT. WMT Unit-II was re-established as an independent unit in 169,170 ha in the Bonggo district of Sarmi, Papua Province, through the issuance of Act No. SK625/Menhut-II/2012.

It was reported that when starting the second period of logging activity in 2012, PT. WMT-II was composed of 46% (77,885 ha) primary forest, 49% (82,902 ha) logged over forest area, and 5% (8,383 ha) swamp forest, open area, and unidentified area (RKU PT. WMT, 2012).

According to the company's work plan (2012-2021), the primary forest harbours 46 commercial tree species, but no information about non-commercial tree data is available. There are three groups of commercial species. The first is fancy wood species, including *Albizia falcataria, Dracontomelum edule, Diospyros* sp, *Podocarpus sp,* and *Pterocarpus indicus.* The second is valuable commercial tree species dominated by *Intsia bijuga, Pometia* sp, *Homalium foetidum, Canarium* sp, and *Celtis rigescens,* among the other 11 species. The third is mixed commercial species comprised of 30 species with *Syzygium* sp, *Pterygota horsfieldii, Terminalia* sp, *Litsea ladermanni,* and *Myristica* sp as the dominant species.

It was also stated that valuable commercial species had the highest harvestable tree density (DBH > 50 cm) of 14.28 stems ha^{-1}, followed by mixed commercial species (11.80 stems ha^{-1}) and fancy wood species (1.28 stems ha^{-1}). In comparison, future trees were dominated by mixed commercial species (71.59 stems ha^{-1}) and valuable commercial species (66.68 stems ha^{-1}). The density of future trees of fancy wood species was significantly lower (5.93 stems ha^{-1}).

2.4. Study plot establishment, arrangement, and design

The research sites were located in primary forest stands and logged forest stands in the concession areas of PT. TTL in Tunas and PT. WMT in Bonggo (Figure 2.1). The study plots were distributed in what were considered representative examples of two major landscape units present in both concession areas; that are: a) fairly flat slope (2-8%) and b) undulating gently steep slope (9-15%). However, the selected logged stands that were intended for 4 and 8 years following harvesting were only available in the steep landscape area of Tunas and Bonggo at the time of data collection.

Therefore, 16 1-ha plots were established in all primary forest stands (four 1-ha plots at every respective landscape units); namely Tunas flat (TF), Tunas steep, Bonggo flat (BF) and Bonggo steep (BS) stands. On the other hand, eight 1-ha plots were evenly laid in four logged stands which were left after logging for four and eight years. These stands were Tunas 4-year, Tunas 8-year, Bonggo 4-year, and Bonggo 8-year logged stand. The term "unlogged forest stand" found elsewhere in this book refers to the primary forest stand as well as its sampling plots.

A 1-ha plot consisted of 10 subplots (20m x 50m), which were arranged in four transect lines (Figure 2.6). The transect lines and subplots were set up as follows: First, the starting point was randomly placed such that the study plots covered the forest stand with a minimum distance of 200m from the stand edge. Then, the first 250m-transect line was run in the East-West direction, and at the same time, three subplots were systematically placed along the line with an in-between space of 50m. After that, the rest of the transect lines were constructed 100m apart from and parallel to each other. Last, when running the lines, three subplots were established in the first- and third-transects, whereas two subplots were in the second- and fourth-transects, which subsequently started at 50m point.

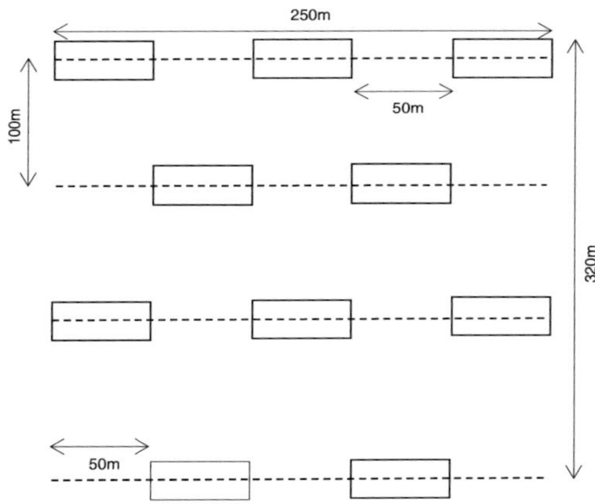

The rectangle indicates subplots (20m x 50m), and the dashed line represents transect lines.

Figure 2.6. Arrangement of subplots.

Since measurements were conducted for different vegetation stages, the subplots (measuring plot) was divided into four compartments. They were called compartment-I to -IV (C-I to C-IV). C-I, with a size of 20m x 50m, bordered the measuring plot. C-II (10m x 25m) was at the centre of C-I, and its long centre line lay along the transect line. Two C-IIIs (5m x 5m) were placed at two different diagonal corners of C-II. One C-IV (1m x 1m) was finally placed at the centre of C-III. The design of the measuring plots is presented in Figure 2.7.

2.5. Data Collection

2.5.1. Measurement in subplots (compartments)

Data explaining the stand's vegetation variables and soil properties were collected in measuring subplots (Figure 2.7). The following measurements were conducted in each compartment:

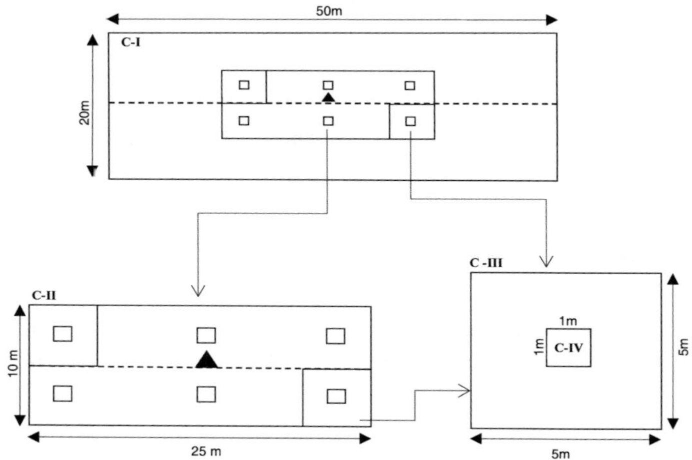

Measurements were conducted for all trees with DBH ≥30cm in compartment I, all tree with 30cm>DBH≥10cm in compartment II, all tree with DBH<10cm and height>1.3m in compartment III, all tree with height<1.3m in compartment IV. Soil samples were taken in triangle, 0-10cm, and 30-40cm depth. C I – C IV: compartment I – IV. The Dash line indicates the centre line of the Transect.

Figure 2.7. Design of measuring plot.

In Compartment-I (20m x 50m), all trees with a diameter at breast height (DBH) equal to and greater than 30cm were marked with paint and numbered. Their species' names, DBH, and height were measured and recorded for data analysis. Tree species were directly identified with the assistance of local experts. In the case of unidentifiable species, specimens (e.g., parts of the tree, such as the leaf, bark, flower, and fruit) were photographed and carried away for further identification at Herbarium Manokwarience of UNIPA, Manokwari, West Papua. Also, DBH of all marked trees was measured at the height of 1.3m above the ground level using phi-band (2-m long tape scaling in diameter). If the wound or buttress was found at that height, then the measurement line would be moved to 2cm above. Moreover, the total tree height was taken by a haga from a fixed distance to the tree.

In Compartment-II (10m x 25m), the same procedure as in C-I was conducted to record species name, DBH, and height of all trees which had DBH greater than or equal to 10cm and less than 30cm.

Natural regeneration was then identified and counted in the next compartments (C-III and C-IV). The regeneration was classified according to the height of the young individuals. Woody stems with total height > 1.3m, and DBH < 10cm were grouped as "sapling". At the same time, those with a total height ≤ of 1.3m were called "seedling" (Lamprecht, 1989).

In Compartment-III (5m x 5m), species' name, diameter, and height of all saplings were measured and recorded using the same procedure as a tree. Whereas, In Compartment-IV (1m x 1m), All seedlings were identified, and their number per species were recorded. In addition, the availability of shrubs, including dominant species, average height, and covering ratio, were recorded in this plot.

In the triangle, soil samples were taken from two different depths, 0 - 10cm and 30 -40cm, using soil rings. The soils were kept in the rings, which were numbered and put in sealed plastic bags when transported to the laboratory. Soil physical characteristics were then identified in the Soil Laboratory of UNIPA Manokwari-

Indonesia, and soil chemical properties were analyzed in the Soil Department of Bogor Agriculture University in Bogor, Indonesia.

2.5.2. Wood sample collection and preparation for nitrogen analysis

The study analyzed and determined the nitrogen (N) content and Nitrogen isotope composition ($\delta^{13}N$) in wood tissue from 50 tree species collected in the Tunas forest and five species taken from the Bonggo forest. 0.5 cm-diameter sapwood samples were extracted using an increment borer from three individuals of each species at 1.3m above the ground on the eastern side and perpendicular to the stem's tangential plane. They were then put in a sealed glass tube to prevent breakage, fungi, or other attacks by micro-organisms and unexpected deterioration. The sample-filled tubes were then labelled with their respective species and replication. In the Wood Science Laboratory of UNIPA Manokwari, they were air-dried to protect the samples from further wood attacks during transportation.

Sapwood samples needed to be prepared in the form of wood powder before proceeding to isotope analysis. The preparation started with drying the wood samples at a constant temperature of 60°C in a Memmert oven for 48 hours to a minimum moisture content of about 28%. After reaching the necessary drying condition, they were finely ground using a ball mill (Micromot 40-E, University of Goettingen, Germany). Prepared wood powders were finally marked and temporarily stored in laboratory containers to hinder water vapour absorption from surrounding air.

About 10-mg of wood powder of each sample was weighed very carefully using a Sartorius M2P electronic microbalance with a computer interface and then placed into a 5 x 9 mm tin capsule. Two standard amounts of a substance similar to the analyzed samples were prepared for each of the 20 wood samples to control the results yielded from the analysis process and monitor the measurement's reproducibility and calibration. During sample preparation, natural abundance and labelled samples were kept and delivered in separate sample wells. Samples were weighed in tin capsules and delivered in 96 well culture trays by ensuring that the culture wells were tightly closed to avoid mixing up samples. Finally, the tin capsules were tightly sealed and formed into small balls to avoid contamination and becoming stuck in the auto sample (KZSI, Unpublished document).

After the preparation, all wood samples were analyzed at the Centre for Stable Isotope Analysis and Research (KIOS), University of Goettingen. The study used a Finnigan MAT stable isotope spectrometer (Delta-S, Finnigan MAT, Bremen, Germany) during the analysis process to measure the nitrogen isotopic signature of woody tissues. The spectrometer is equipped with a Hereaus element analyzer (EA-1100, Carlo Erba, Milan, Italy) for combustion samples under an excess of oxygen via a ConFlo III interface.

According to Dyckmans (Unpublished document), an Elemental Analyzer (EA) consists of an auto sample, an oxidation reactor, a reduction reactor, a water trap, and a gas chromatograph (GC) column in which the oxidation chamber is set at 940°C. The column contains tungsten and silvered cobaltous oxides for the combustion process, while the reduction tube is set at 600°C and contains copper granules. The water trap contains magnesium perchlorate.

During the analysis process, the sample is converted to NO_x by flash combustion taking place in the oxidation reactor. As a result, the N-containing combustion gases are reduced in the reduction reactor to N_2 with copper as a catalyst. Sample gases pass a water trap and are then separated by GC and analyzed in the mass spectrometer (MS). In isotope ratio mass spectrometry (IRMS), heavy and light isotopes are detected in parallel in different detector cups. The isotope abundance is calculated from the number of molecules trapped in the Faraday cups (Dyckmans, Unpublished document).

Since variation in the absolute abundance of ^{15}N is small, nitrogen isotope composition is expressed in delta notation and given as:

$$\delta^{15}N(^{o}/_{oo}) = \left(\frac{R_{sample}}{R_{standard}} - 1 \times 1000\right)$$

In which, R_{sample} and $R_{standard}$ are ratios of heavy to light isotopes ($^{15}N/^{14}N$) of the sample and the atmospheric N_2 standard, respectively (Koopmans *et al.*, 1997; Koba *et al.*, 2003; Domingues *et al.*, 2007). The value of the $R_{standard}$ used in the calculation is 0.0036765 (Evans, 2001), and the precision of measures is 0.2 ‰ (Koopmans *et al.*, 1997; Guehl *et al.*, 1998; Domingues *et al.*, 2007).

Figure 2.8. Field work

2.6. Data analysis

Species area curves were first constructed based on Cain's method (Cain, 1938) to evaluate the acceptable total sample's area. The standard error of the mean basal area of sample plots in each sampling site was calculated to indicate the accuracy and precision of the measured data. Tree compositions and important value index (IVI), a relative abundance, relative frequency, and relative dominance sum-up at species and family level, were then calculated in tabular format, using general procedures and equations developed by Curtis & Mcintosh (1950) and Lamprecht (1989). Moreover, species diversity was determined using the Shannon-Wiener index and Shannon measure of evenness, while species similarity was identified using Sorensen's coefficient, as suggested by Magurran (1988) and Lamprecht (1989).

In examining forest structure, the study calculated diameter variables (stand's mean-, arithmetic-, and quadratic-diameters, and basal area) and height variables (mean height, vertical structure, 20% thickest stems height) following Lamprecht (1989) and van Laar & Akça (2007) and presented in tabular form. The calculation also included a height–diameter curve and tree-diameter distribution, shown in a graph form. In addition to testing hypotheses, the study used inferential statistics when comparing structural parameters. Parametric and nonparametric statistical analyses, such as ANOVA and the Kruskal Wallis test, were applied to test the significant differences among forest structure variables.

Regression analysis is a conceptually simple method for investigating functional relationships among variables (Chatterjee & Hadi, 2015). In this study, simple and multiple regression analyses were used to find a relationship and correlation between forest variables (height- DBH and diameter classes) to construct models based on forest variables (Vanclay, 1994; van Laar & Akça, 2007).

A more detailed explanation of the methods and formulas needed are included in the relevant chapters. All tabular and statistical analyses were conducted in Microsoft Excel 2010, Estimate S version 9, and Statistica software version 11.

3. Accuracy and precision of sample data

3.1. Data accuracy

The sample unit size and standard error of estimated means were evaluated using qualitative (tree species) and quantitative (tree basal-area/BA) attributes to show the balance between quality and efficiency of sampling design in this study. The deviation between the number of species sampled and the exact number of species in the respective population explains how closely the sample size represents the whole stand (accuracy) that can be simply expressed by a species accumulation curve. On the other hand, the variation of sampled tree BAs around their mean indicates the preciseness of measuring data that is shown as percent standard error.

The species accumulation curve has been widely used in tree species composition and diversity studies (Lamprecht, 1989; Pham, 2012; Khaing, 2013; Cam, 2015) to indicate an acceptable sample unit (individual- or area-based) size in a study area. The accumulation curve records the addition of species collection as individuals or areas are added to the pool of all previously observed samples. When the sample unit is continuously enlarged, asymptotic (a maximum number of species collected) will be reached if no additional species are encountered. The minimum size of the sample unit that corresponds to the curve asymptote is considered the efficient size of acceptable representative of the population. Statistically, the minimum sampling unit size is agreed upon when new species' occurrence is 10% per 10% sample size enlargement (Cain, 1938; Lamprecht, 1989).

Species accumulation curves for four primary forest stands and four logged forest stands are presented in Figure 3.1.

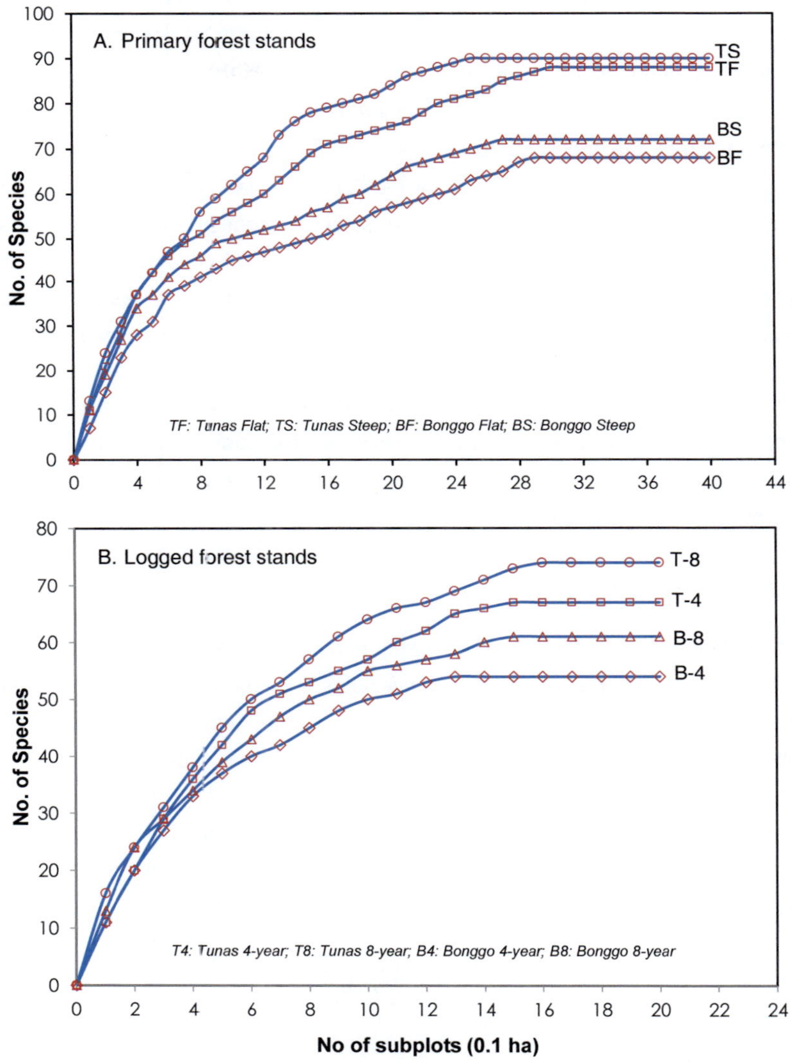

Figure 3.1. Species accumulation curves for primary forest stands (A) and logged forest stands (B) of Tunas and Bonggo. The species name was recorded from all trees with DBH ≥ 10 cm.

The graphs show that the maximum registered species varied largely among plots (see the curves' asymptote), but these expected total species of all the respective stand's populations could be reached within the sampling plot areas of 4.0ha primary forests, and 2.0ha logged forest applied in this study. In the primary forest, the largest sampling plot size of 3.0ha to ensure the accuracy in registering species was represented by the Tunas flat stand and the smallest size plot of 2.5ha was shown in the Tunas steep stand. In contrast, the logged forests needed a smaller sampling plot size than those of primary forests to confirm all species were investigated; 1.6ha (the largest) and 1.4ha (the smallest) sampling plot size were indicated by Tunas 8-year and Bonggo 4-year stand, respectively.

The result implied that the sampling method applied in this study was highly accurate in terms of species number. However, Richards (1996) noted that in some primary tropical lowland forests, a species-area curve can reach its asymptote after 4.0 to 5.0ha sampled areas.

3.2. Data precision

The standard error of the mean basal area of sample plots in a sampling design is mostly considered as an indicator in evaluating the precision of measured data before further data analysis in forest structure studies (Lamprecht, 1989). The allowable percent standard error is less than 10% (Zöhrer, 1980).

Basal areas of trees with DBH≥10 cm were measured in 40 subplots of primary forest stands and 20 subplots of logged forest stands. Then, the sample-based standard error of the mean of each stand was estimated as follows:

$$S_{\bar{g}} = \sqrt{\frac{\Sigma(g_i-\bar{g})^2}{n-1}} \qquad 3.1$$

In estimating the relative accuracy of the recorded data, the standard error of the mean basal area in percentage was applied using the following formula:

$$S_{\bar{g}}\% = \frac{S_{\bar{g}}}{\bar{g}} \, x \, 100 \hspace{4cm} 3.2$$

Where: $S_{\bar{g}}$ = Standard deviation from the mean basal area of subplots
g_i = Basal area of ith subplot
$\bar{g}$ = Arithmetic mean basal area of subplot
n = number of subplots

Standard errors of the mean basal areas (in percentage) for all study plots are presented in Table 3.1 (primary forests) and Table 3.2 (logged forests).

Table 3.1.   Mean basal area and standard error (in absolute and percentage terms) calculated from 40 0.1-ha subplots in the primary forest of Tunas flat (TF), Tunas steep TS), Bonggo flat (BF), and Bonggo steep (BS).

Stands	TF	TS	BF	BS
No of Subplots (n)	40	40	40	40
Mean Basal Area (BA) ($m^2/0.1ha$)	2.145	2.540	2.721	3.456
Standar Error (SE)	0.158	0.175	0.260	0.276
Percent of Standar Error (%)	7.4	6.9	9.6	8.0

The percent standard errors of basal areas of all stands in intact forests (Table 3.1) were less than 10% in all plots; 7.4%, 6.9%, 9.6%, and 8.0% from TF, TS, BF, and BS, respectively. On the other hand, not all the collected data from stands in the logged forests (Table 3.2) were precise in measurement (%SE of T4 > 10%). The study noted most big gaps in stand T4 had just been covered mostly by saplings than trees.

Table 3.2.   Mean basal area and standard error (in absolute and percentage terms) calculated from 20 0.1ha subplots in Tunas 4-year logged (T4), Tunas 8-year logged (T8), Bonggo 4-year logged (B4), and Bonggo 8-year logged stand.

Stands	T4	T8	B4	B8
No of Subplots (n)	20	20	20	20
Mean Basal Area (BA) ($m^2/0.1ha$)	0.892	1.018	0.853	1.139
Standar Error (SE)	0.101	0.100	0.078	0.097
Percent of Standar Error (%)	11.3	9.8	9.1	8.6

It can be inferred from the accuracy of sampling measurement that registered data in all stands were adequate for estimating the species composition, diversity and forest structure characteristics. Even though the precision of measurement in stand T4 was relatively low, the data could be included for further analysis as the %SE of 11.32% remained close to the lower limit of 10%.

4. Intact primary forests of Tunas and Bonggo

4.1. Tree family and species composition

4.1.1. Family composition

This chapter reports tree families and species of untouched stands at logging concessions in Tunas and Bonggo forests and analyzes the important values (IV) of family and species in respective stands. Based on the analysis, the study determined to what extent the IVs explain the distribution and establishment of the tree families and species locally (within the forest) and regionally (between forests), which vary in physiographic setting and soil types.

Stand selection and establishment and tree sample collection, identification, and measurements, are presented in Chapter 2. The study recorded all trees with DBH > 10cm from a 4-ha plot in every individual stand of Tunas Flat (TF), Tunas Steep (TS), Bonggo Flat (BF), and Bonggo Steep (BS). The tree families and species were then ranked based on their Important Values (IV), which were expressed in an index of family importance value (FIV) (Mori *et al.*, 1983) and species' important value (IVI) (Curtis & Mcintosh, 1950). The indices are a combination of relative dominance, relative abundance, and relative frequency of all plant families and species in a stand; the calculation of the indices is as follows:

$$\text{FIV or IVI (\%)} = RA + RF + RD \qquad 4.1$$

Where:

F/IVI	=	Important Value Index of a family or a species
RA.	=	Relative abundance (%) of a family or species reflects a proportion of its individuals compared to the total individuals in a study stand.
RF.	=	Relative frequency (%) is a ratio of plots occupied by a family or species and total plots in a study stand.
RD.	=	Relative Dominant (%) is a family or species' basal area percentage from a specified area in a study stand.

This study identified 43 families, and every family was composed of 1 to 11 species that made up the 143 species encountered in the Tunas and Bonggo sites. Table 4.1 displays tree families that had three or more species and their distribution in four sampled stands (see also Appendix 1 for detail). Those families were called "common families," and the rest were named "rare families."

Table 4.1. Common tree families (DBH > 10 cm) with three and more species encountered in the primary moist forest of Tunas and Bonggo, Papua

			No. of species found in			
		No. of all Species	Tunas		Bonggo	
No	Families		Flat	Steep	Flat	Steep
1	Burseraceae	11	7	8	6	7
2	Myristicaceae	10	9	5	6	6
3	Lauraceae	9	7	8	2	2
4	Meliaceae	9	6	5	4	5
5	Sapotaceae	9	8	6	1	2
6	Myrtaceae	8	7	7	5	5
7	Anacardiaceae	7	2	1	5	6
8	Euphorbiaceae	7	5	6	3	3
9	Clusiaceae	6	5	6	3	3
10	Fabaceae	6	-	1	5	4
11	Rubiaceae	6	2	2	3	3
12	Dipterocarpaceae	4	4	4	2	4
13	Elaeocarpaceae	3	2	2	2	2
14	Lamiaceae	3	2	2	1	1
15	Malvaceae	3	1	1	2	2
16	Moraceae	3	-	-	2	3
17	Sapindaceae	3	2	3	2	2
	Other families (26)	36	18 (17)	23 (22)	14 (14)	12 (12)
	Total families (43)	143	87 (32)	90 (38)	68 (31)	72 (29)

The number of families is indicated in the parenthesis.

Most of the common-families were found in all four studied stands. They varied in species number, and none of them had species in all stands. Common families with a higher number of species, such as Burseraceae (11 species), Myristicaceae (10 species), and Meliaceae (9 species), distributed their species evenly (5 to 9 species) among the stands; however, they shared less than half of their species between Tunas and Bonggo sites. Other higher-species families, for instance, Lauraceae (9 species) and Sapotaceae (9 species), were imbalanced in their species distribution. Both families were concentrated more on Tunas plots; Tunas stands were composed of eight species of these families, compared to two species in the Bonggo stands. In other cases, some common-families with fewer species shared all their species between the two forests, i.e., Dipterocarpaceae (4 species), Elaeocarpaceae (3 species), and Sapindaceae (3 species), but not Fabaceae (6 species) and Moraceae (3 species), which were uncommon in the Tunas forest.

Like common families, some rare families were also recorded both in Tunas and Bonggo forests; the two forests shared 12 out of 26 rare families. These families were, however, represented by different species, either in Tunas or Bonggo stands. Apocynaceae, for example, was represented by *Alstonia spectabilis* in Tunas but *Alstonia scholaris* in Bonggo. Other cases were Combretaceae (*Terminalia arborea* vs. *Terminalia* sp) and Thymelaeaceae (*Aquilaria polyantha* vs. *Aquilaria* sp). Most of the species composed of these 12 shared families were the canopy species, i.e., Thymelaeaceae. Further to the other 14 rare families, the Tunas forest markedly composed more unshared families than the Bonggo forest, respectively 12 versus two families. Middle-storey species mostly belong to these unshared rare-families, e.g., *Gonocaryum tiriform* (Cardiopteridaceae), *Lithocarpus rufovillosus* (Fagaceae), *Maniltoa plurijuga* (Leguminaceae), and others.

The tree taxa distribution implies that the two forests have high similarity at the family level. However, there is a gradual increase in taxa distribution restriction from higher-order (family) to the lower-order (species). There is a smaller number of taxa in the Bonggo plots; perhaps the forest is more isolated by mountains than the Tunas forest.

Table 4.2 shows ten dominant families ranked by their FIVs in flat and steep stands of Tunas' primary forest. All the families had four or more species (common-families), and they highly dominated the stands, as indicated by the total FIVs of 236.86% for flat stand and 231.61% for the steep stand.

Dipterocarpaceae held a central position at the flat stand. It had the highest number of individuals (53 stems ha^{-1}), the most frequent species (100%), and the largest basal area (5.25 m^2.ha^{-1}), which altogether resulted in the highest FIV of 53.13%. The following essential families were Lauraceae, Myrtaceae, and Euphorbiaceae, with FIV of 31.01%, 28.89%, and 23.60%, respectively. The Dipterocarpaceae still played the primary role at the steep stand, and showed practically no differences in absolute abundance (only two stem ha^{-1} more), frequency (only 2% less), and dominance (only 0.80 m^2.ha^{-1} less). However, its FIV was significantly reduced to 45.35% in the steep stand. The FIV reduction of the Dipterocarpaceae was most likely because of the other ten top-ranked families' increasing importance. Burseraceae contributed the most. Its individuals increased the abundance (to 15 stem ha^{-1} more), the frequency (to 2% more), and the dominance (to 7.12 m^2.ha^{-1} more), which upgraded its rank from fifth place with a FIV of 23.21% at the flat stand to second place, with a FIV of 32.20%, at the steep forest. Next to Burseraceae was Myrtaceae; there was no increase in individuals, but Myrtaceae represented significantly larger stems in the steep stand. Other common families, such as Lauraceae, Euphorbiaceae, and Clusiaceae, made a small contribution to the reduction because they presented an increase only in absolute abundance, and neither in absolute basal area nor absolute frequency. It implies more juveniles (small-sized individuals) from existing and additional species of these families in the steep stand. In short, even though Dipterocarpaceae was predominant in Tunas plots, Burseraceae and Myrtaceae showed a more substantial development in the steep stand; they showed a significantly higher number of big-sized trees in the stand compared to other families.

Table 4.2. The ten most important families (DBH>10cm) ranked based on Family Important Value (FIV) at two topographic stands in **Tunas** primary forest.

No.	Family	Absolute			FIV (%)
		Abundance (N.ha^{-1})	Dominance (m^2.ha^{-1})	Frequency (%)	
	Flat forest stand				
1	Dipterocarpaceae	53	5.25	100	53.13
2	Lauraceae	32	2.16	100	31.01
3	Myrtaceae	26	2.39	93	28.89
4	Euphorbiaceae	22	1.70	85	23.60
5	Burseraceae	23	1.60	83	23.21
6	Sapotaceae	18	1.29	83	20.17
7	Meliaceae	18	1.27	68	18.53
8	Myristicaceae	14	0.84	68	15.16
9	Lamiaceae	9	0.93	55	12.70
10	Clusiaceae	8	0.63	50	10.46
	Other families (22)	46	3.40		63.14
	Total (32)	269	21.45		300
	Steep forest stand				
1	Dipterocarpaceae	55	5.12	98	45.35
2	Burseraceae	38	3.38	88	32.20
3	Lauraceae	35	2.14	100	27.55
4	Myrtaceae	24	3.10	90	26.82
5	Euphorbiaceae	35	1.84	93	25.61
6	Meliaceae	20	1.50	93	19.64
7	Sapotaceae	15	1.30	73	15.66
8	Clusiaceae	14	1.00	80	14.79
9	Myristicaceae	16	1.00	63	14.09
10	Lamiaceae	10	0.71	50	9.90
	Other families (28)	58	4.31		68.39
	Total (38)	319	25.40		300

Table 4.3. The ten most important families (DBH>10cm) ranked based on Family Important Value (FIV) at two topographic stands in **Bonggo** primary forest.

No.	Family	Absolute			FVI (%)
		Abundance (N.ha^{-1})	Dominance (m^2.ha^{-1})	Frequency (%)	
	Flat forest stand				
1	Fabaceae	24	5.19	73	32.07
2	Myristicaceae	47	1.94	80	26.60
3	Salicaceae	43	1.63	83	24.70
4	Myrtaceae	36	1.59	73	21.91
5	Burseraceae	36	1.51	68	20.98
6	Dipterocarpaceae	27	1.74	73	20.28
7	Sapindaceae	20	2.14	63	18.91
8	Cannabaceae	36	1.16	58	18.74
9	Euphorbiaceae	24	1.48	53	16.48
10	Anacardiaceae	16	1.45	45	13.79
	Other families (21)	95	7.39		85.55
	Total (31)	402	27.21		300
	Steep forest stand				
1	Fabaceae	24	6.49	88	30.75
2	Myristicaceae	61	2.22	80	25.23
3	Myrtaceae	50	2.38	75	23.06
4	Dipterocarpaceae	33	2.78	78	20.97
5	Burseraceae	44	1.93	80	20.94
6	Sapindaceae	29	2.79	83	20.53
7	Cannabaceae	37	1.76	75	18.74
8	Salicaceae	24	1.84	65	15.37
9	Lauraceae	27	1.47	63	14.76
10	Euphorbiaceae	25	1.20	60	13.35
	Other families (19)	142	9.70		96.32
	Total (29)	493	34.56		300

The ten most prominent families of flat- and steep-slope stands in the Bonggo primary forest ranked by their FIVs are illustrated in Table 4.3. Unlike those in the Tunas forest, in which all ten were common-families, some of the dominant families in the Bonggo forest had single species (rare-families); these were Salicaceae and Cannabaceae. Table 4.3 also shows that the ten-top families markedly dominated the flat and steep stands. The sums of FIVs of the ten families were 214.45% in the flat stand and 203.68% in the steep stand. They are about 72% and 68% of the total FIVs of the respective stand.

Fabaceae was the most important family at the flat and steep stands of the Bonggo primary forest; it had the highest FIV (32.07% and 30.75% at respective stands). It was exceptional that Fabaceae, which was the least-abundant (24 individuals ha^{-1}) family, occupied the largest basal-area at both stands; the family's dominance values were markedly higher (5.19 m^2.ha^{-1} at the flat stand and 6.49 m^2.ha^{-1} at the steep stand) than any other top families. In contrast, Myristicaceae, as the second-most important family, had a higher individual count of 47, 61 stem ha^{-1}, and very low absolute dominance of 1.94 m^{2}ha^{-1} and 2.22 m^{2}ha^{-1} in the respective stands. The rest of high-importance families at the lower level showed the same pattern as the Myristicaceae; many in individuals (high abundance) even significantly had more abundance in the steep stand but were represented by less big-sized trees (low in dominance).

The tree composition of Bonggo stands explained that Fabaceae predominantly inhibited the stands and markedly suppressed the other families, which had less chance to grow to large individuals or emerge as canopy trees, which they are capable of doing under some circumstances; for example, the individuals of Dipterocarpaceae, Burseraceae, and Myrtaceae can grow to large-sized trees, as shown by their diameter size in Tunas stands. The higher abundance of other important-families that mainly occupied the middle and lower canopy also showed Fabaceae's suppression. This was reflected in the higher stem density at the steep and flat stands (493 and 402 stems ha^{-1}), which were significantly higher than those of Tunas stands (269 and 319 stems ha^{-1}).

4.1.2. Species composition

The top ten important tree species (DBH >10 cm) recorded in flat and steep stands of Tunas primary forest are presented and ranked based on the Important Value Indices (IVIs) in Table 4.4. The species' amounted indices were 123.28% and 113.90% (about 41% and 38% of the total IVI 300%). In both stands, there were a total of 12 important species in six families: *Vatica rassak, Anisoptera polyandra, Hopea papuana* (Dipterocarpaceae), *Litsea timoriana, Beilsmiedia* sp (Lauraceae), *Blumeodendron amboinicum, Pimelodendron amboinicum, Medusanthera polot* (Euphorbiaceae), *Canarium asperum, Dacryodes* sp (Burseraceae), *Syzygium versteegii, Syzygium anomalum* (Myrtaceae), and *Teijsmanniodendron bogoriense* (Lamiaceae).

Vatica rassak, the highest among the ten most dominant species in both stands, reached an IVI of 35.80% and 31.93% in flat and steep stands, respectively. These IVIs were more than twice those of the IVI of the second-most important species, *Litsea timoriana,* which were 16.33% in the flat plot and 14.47% in the steep plot. The two highest-importance species had remarkedly high abundances (38 and 17 individuals ha^{-1}), and they were distributed in almost all plots of the two stands (relative frequency above 83%). In contrast, other high-importance species were only represented by 7 to 14 individuals ha^{-1} and scattered among less than 55% of the plots in the stands.

There were slight differences in the species composition between the flat and steep stands. *Vatica rassak* and *Litsea timoriana* showed greater absolute abundance, frequency, and dominance in the steep stand. However, their relative importance compared to other species in that stand was lower than their IVIs in the flat stand. Other important species, such as *Anisoptera polyandra, Hopea papuana, Blumeodendron amboinicum, Teijsmanniodendron bogoriense, Pimelodendron amboinicum*, and *Beilschmiedia* sp, also had lower IVIs in the steep stand, with only a slight change in their absolute values. On the other hand, *Canarium asperum and Dacryodes* sp demonstrated a significant increase in their IVIs, due to the appearance of bigger individuals in the steep stand; a significantly higher absolute number of individuals and dominance reflected this increase. It was the main reason why the dominance of *Vatica rassak* and *Litsea timoriana* was reduced in the steep stand. Increasing stem numbers of existing and incoming rare species also most likely affected the dominance of Vatica rassak and Litsea timoriana and the other top important species if a comparison of species IVs is made for the flat and steep stands.

In the case of the Bonggo primary forest, Table 4.5 shows that the summed IVI of the ten most important species reached more than 50% of the total stand's IVI (approximately 152.94%) in the flat stand, but the number was slightly less for the steep stand (138.95%). They were composed of 12 species in ten families from a total of two stands.   The leading cause of high proportional values of the dominant species was that some of them had many emerging big trees with a fewer abundance of small stems; examples of this situation included *Intsia bijuga* (Fabaceae), *Anisoptera polyandra* (Dipterocarpaceae), and *Pometia pinata* (Sapindaceae). The other species with a high abundance of small- to middle-size trees were *Homalium foetidum* (Salicaceae), *Celtis rigescens* (Canabaceae), *Myristica tubiflora*, *Myristica sulcata* (Myristicaceae), *Syzygium* sp, *Syzygium anomalum* (Myrtaceae), *Litsea ledermannii* (Lauraceae), *Pimelodendron amboinicum* (Euphorbiaceae), and *Canarium asperum* (Burseraceae).

Homalium foetidum was the most important-species in the flat stand, followed by *Intsia bijuga*. However, in the steep stand, the importance of *Intsia bijuga* was greater than that of *Homalium foetidum,* which was demoted to second place. With fewer individuals (24 stem ha-1) and only a slight increase in basal area (0.21 m2ha-1) in the steep stand, the IVI of *Homalium foetidum* increased insignificantly.

Table 4.4.   The ten most important tree species with DBH>10cm ranked based on their Important Value Index (IVI) of two different topographic conditions (flat and steep forest stands) at **Tunas** primary forest.

No.	Species	Absolute			IVI (%)
		Abundance $(N.ha^{-1})$	Dominance $(m^2.ha^{-1})$	Frequency (%)	
	Flat forest stand				
1	*Vatica rassak*	38	3.47	93	35.80
2	*Litsea timoriana*	17	1.05	85	16.33
3	*Blumeodendron amboinicum*	10	0.77	55	10.41
4	*Anisoptera polyandra*	7	1.04	43	10.02
5	*Teijsmanniodendron bogoriense*	8	0.85	48	9.90
6	*Pimelodendron amboinicum*	8	0.59	48	8.70
7	*Canarium asperum*	8	0.57	45	8.41
8	*Hopea papuana*	7	0.68	45	8.38
9	*Beilschmiedia* sp	7	0.58	48	8.08
10	*Syzygium versteegii*	7	0.46	40	7.28
	Other species (77)	152	11.39		176.72
	Total species (87)	269	21.45		300
	Steep forest stand				
1	*Vatica rassak*	40	3.64	95	31.93
2	*Litsea timoriana*	18	1.10	83	14.47
3	*Canarium asperum*	13	1.18	55	11.59
4	*Blumeodendron amboinicum*	11	0.69	60	9.38
5	*Dacryodes* sp	9	1.14	30	8.74
6	*Hopea papuana*	8	0.76	53	8.24
7	*Medusanthera polot*	11	0.56	48	8.11
8	*Syzygium versteegii*	7	0.72	50	7.69
9	*Syzygium anomalum*	7	0.66	43	6.98
10	*Teijsmanniodendron bogoriense*	8	0.56	40	6.80
	Other species (80)	189	14.39		186.10
	Total species (90)	319	25.40		300

Table 4.5. The ten most important tree species (IVI rank) with DBH>10cm found in two different topographic conditions (flat and steep forest stands) at **Bonggo** primary forest.

No.	Species	Absolute			IVI (%)
		Abundance (N.ha⁻¹)	Dominance (m².ha⁻¹)	Frequency (%)	

No.	Species	Abundance $(N.ha^{-1})$	Dominance $(m^2.ha^{-1})$	Frequency (%)	IVI (%)
	Flat forest stand				
1	*Homalium foetidum*	43	1.63	83	23.52
2	*Intsia bijuga*	13	4.04	55	22.67
3	*Celtis rigescens*	36	1.16	58	17.91
4	*Anisoptera polyandra*	23	1.51	60	16.35
5	*Pometia pinnata*	18	1.74	53	15.19
6	*Myristica tubiflora*	25	1.23	53	15.00
7	*Syzygium* sp	19	0.91	43	11.64
8	*Litsea ledermannii*	15	1.01	48	11.37
9	*Pimelodendron amboinicum*	15	0.71	43	9.82
10	*Myristica sulcata*	17	0.55	40	9.46
	Other species (58)	180	12.73		147.06
	Total species (68)	402	27.21		300
	Steep forest stand				
1	*Intsia bijuga*	14	5.49	80	23.88
2	*Celtis rigescens*	37	1.76	75	17.52
3	*Anisoptera polyandra*	27	2.26	70	16.55
4	*Homalium foetidum*	24	1.84	65	14.32
5	*Pometia pinnata*	21	1.93	65	14.07
6	*Litsea ledermannii*	24	1.31	63	12.60
7	*Myristica sulcata*	26	1.07	55	11.90
8	*Syzygium anomalum*	24	1.09	55	11.53
9	*Pimelodendron amboinicum*	17	0.87	38	8.46
10	*Canarium asperum*	16	0.71	43	8.11
	Other species (62)	264	16.22		161.05
	Total species (72)	493	34.56		300

The third-place species in the flat area was *Celtis rigescens*, with an IVI of 17.91, followed by *Anisoptera polyandra* (IVI 16.35%) and *Pometia pinnata* (IVI 15.19%) in the fourth- and fifth places, respectively. In the steep stand, these five species were ranked the same as in the flat area, except for the downgrading of *H. foetidum* from the first to the fourth place.

There was a considerably higher number of stems (abundance) and basal areas (dominance) in the steep stand than in the flat stand. The less-important species represented 84 out of 91 trees (92%), which were the difference stems between the two stands, whereas only seven trees (8%) were distributed among the ten most-important species of the steep stand. On the other hand, the ten most-important species covered 3,86 out of 7.35 m^2ha^{-1} (53%) basal area, with more area covered by the species in the steep stand than in the flat stand. These numbers imply that the presence of more big trees of the ten most-important species in the steep stand caused the difference in basal areas between the two stands. In contrast, there was a remarked increase in the recruitment of small-sized trees from less-important species.

Among the ten most-important species, only *Intsia bijuga* (38%), *Anisoptera polyandra* (20%), and *Pometia pinnata* (5%) contributed the most excessive basal area in the steep stand. The contribution was identical to the added numbers of big-sized trees (DBH > 50 cm). The number of big stems per ha increased respectively from flat to steep stand as follows: 6 to 9 (*Intsia bijuga*), 2 to 3 (*Anisoptera polyandra*), and 1 to 3 (*Pometia pinnata*). Besides the most important species, there are 11 less important species with big trees equal to or fewer than two steems ha^{-1}, such as *Albizia falcataria, Alstonia scholaris, Arthocarpus* spp, *Campnosperma auriculatum, Dillenia alata, Palaquium obtusifolium, Parartocarpus* spp, *Pterocarpus indicus, Pterygota horsfieldii, Swietenia macrophylla, Terminalia arborea,* and *Toona sureni.*

4.1.3. Stand composition at another site in western New Guinea

Here, the tree species compositions from Tunas and Bonggo sites are compared to those from another lowland forest in West New Guinea. Wagner (2016) investigated tree species (DBH > 10 cm) in 2 1-ha plots of primary forest at a logging concession in Sungguan Valley, northern Bird's head of West Papua. At the Sungguan site, 109 species in 38 families were found. These numbers are relatively close to those listed in the Tunas site (94 species in 40 families) but significantly higher compared to those recorded in the Bonggo site (72 species in 31 families).

Sungguan plots share 30 families with Tunas plots, of which six families (Rutaceae, Phyllanthaceae, Stemonuraceae, Rosaceae, Putranjivaceae, and Cardiopteridaceae) are uncommon in the Bonggo site. These numbers are higher than the 25 families that are in common between Sungguan and Bonggo plots. Only one family (Moraceae) is restricted to the Tunas site. Nevertheless, the three sites share 24 families.

Moreover, about 70% of families commonly found in all sites are those with three or more species in every location. Still, their genera and species vary in individual and number. For example, Moraceae, the most species-rich family (eight species in five genera) in the Sungguan site, is represented by only three species in three different genera in the Bonggo site. This is surprising because the two areas only shared one species of Moraceae (*Ficus variegata*). No Moraceae is in the Tunas site, emphasizing that the distribution is mainly in the northern lowland. The other species-abundant families that were northern-oriented were Anacardiaceae and Fabaceae; they were abundant in Sungguan and Bonggo but species-poor in Tunas.

In contrast to Sungguan-Bonggo oriented families, Lauraceae has more species in Sunguan (four species in two genera) and Tunas (nine species in five genera) in comparison to Bonggo (two species in two genera). Lauraceae also shares all its species between Sungguan and Tunas sites; that is markedly higher than

Lauraceae's shared species (only one - *Litsea ladermanii*) between Sungguan and Bonggo sites. Another family, Burseraceae, is quite different because it only appears in the Bonggo and Tunas sites. Listed as a species-poor family in Sungguan, Burseraceae is, on the contrary, the most species-rich family in Tunas (8 species in 4 genera) and Bonggo (7 species in 2 genera). This is interesting because Burseraceae shares its only single species, *Canarium hirsutum*, between the Sungguan and Tunas sites.

The next-most species-rich families, Euphorbiaceae, Meliaceae, Myristicaceae, and Myrtaceae, have five or more species and are evenly distributed in Sungguan, Tunas, and Bonggo sites.

In terms of community domination, the family-important-indices (FIVs) of the ten top-ranked families vary among Tunas, Bonggo, and Sungguan sites; however, some families indicate similarity in their distribution and dominance. To show dominant families and their closed distribution between the three sites, they can be differentiated into four groups. The first group is the families that are only the most important in both Sungguan and Bonggo plots. Sapindaceae with the highest FIV (31.1%) at Sungguan belongs to this group. At the Bonggo plot, Sapindaceae is in the sixth place in FIV (20.5%); however, it is relatively less important in the Tunas site, with a FIV of 7.3%. Like Sapindaceae, Fabaceae and Anacardiaceae are the other ten most important families at Sungguan that are also crucial in Bonggo; they place first and tenth, respectively. The second group is families high-ranked only in Sungguan and Tunas sites. A single-family represents this group; Sapotaceae is the only exclusive family in the two locations. It ranked tenth (12.9%) at the Sungguan plot and seventh (15.7%) at the Tunas plot. The third group is those that are only important in the Sungguan plot. Annonaceae, with the second-highest FIV (28.3%) at the Sungguan plot, has low FIV at both Bonggo (5.9%) and Tunas (0.6%). Two other families, Moraceae and Malvaceae, show the same exclusive importance as Annonaceae in the Sungguan site, i.e., they are, respectively, the sixth and the seventh most essential families at Sungguan. The last group includes families listed as the most dominant in all three sites, such as Euphorbiaceae, Lauraceae, and Meliaceae.

Most of the families found in this study are pantropically distributed, with many genera and species. For example, Burseraceae has about 550 species in 16 genera, but only five genera of Canariaceae are distributed in Malesia (Leenhouts *et al.*, 1955), and only two genera (Dacriodes and Haplolobus) are listed in this study. Lauraceae includes about 2500 species in 31 genera, with Beilschmeidea, Cryptocarya, Litsea, Actinodaphne, and Cinnamomum having their centre of origin in Malesia and distribute from Asia to Australia (Kostermans, 1957). All these genera are listed in the study sites, including the Sungguan plot. Myristicaceae has about 500 species in 20 genera, from which about 350 species in the genera of Horsfieldia, Knema, and Myristica are distributed between Asia and New Guinea (Wilde, 2000). This study only registered two genera (Horsfieldia and Myristica), but Myristica is the most species-abundant genera. Myristica has a broad distribution from India to North Australia, with the centre of speciation in New Guinea (about 100 from 170 species of Myristica). The taxa distribution indicates that New Guinean lowland forest flora is essentially Malesian, with relatively Australian components, as confirmed by other studies (e.g., Paijmans, 1976; Johns, 1995).

The tectonic collision between the Australian and Southeast Asian plates about 10 million years BP has allowed flora and fauna from Gondwanan and Laurasian to migrate onto New Guinea (Walker & Hope, 1982). Thus, the island's continual uplifting has formed today's physiographic and environmental condition, which has affected the distribution and isolation of modern plants and animals (Walker & Hope, 1982). The variation and similarity of tree family and species composition in the Tunas, Bonggo, and Sungguan sites clearly explain the effect of the island's geological evolution. However, the reason for the taxa's widely spread orientation, including the three distant lowland forest sites in Western New Guinea, cannot be explained without specific further investigation.

4.2. Tree species diversity and similarity in the primary forest

4.2.1. Species richness

A combination of site conditions like soil characteristics, light, slope, water supply, and disturbance is vital in determining the level of species diversity (Magurran, 1988). Tree species diversity in New Guinea varies from moderate in most lowlands (Paijmans, 1970; Johns, 1985; Kiapranis, 1991; Oatham & Beehler, 1995) to high in the highlands (Wright *et al.*, 1997). This is because the island is composed of the combination of original flora from Gondwanan and Laurasian (Walker & Hope, 1982) and experiences rapid speciation as a result of environmental instability and isolation due to the continual uplifting of the island (Wright *et al.*, 1997). Since the southern and the northern lowlands of Papua have been experiencing different geological evolutions (Polhemus, 2007), this study hypothesized that the tree species in Tunas and Bonggo stands, which are located in the two lowlands, have different levels of diversity due to ecological and evolutionary processes. The objectives of the study (in this section) were to determine the level of tree species diversity for each stand with appropriate indices, explain the possible biological process, underline the diversity patterns of the stands, and identify species similarity among stands.

The study involved four study stands located on two different slopes (flat and steep) in each of the Tunas and Bonggo forests, named Tunas flat (TF), Tunas steep (TS), Bonggo flat (BF), and Bonggo steep (BS). Data collected and used for the analysis were species names and the number of all trees with a diameter at breast height (DBH) $\geq$ 10 cm recorded in 40 0.1-ha subplots (Figure 2.5)

Tree species diversity of the four study stands was evaluated at two spatial scales, the plot and stand levels, and analyzed with two indices (Hill, 1973), species richness and species heterogeneity. Species richness was expressed in the mean number of species per plot and extrapolated to stand level. Species heterogeneity was presented in Log Serie α, Shannon (*H'*), Shannon Evenness

(E), Shannon exponential (e^H), and reciprocal of Simpson index (1/D). Species similarity between stands was expressed in a Jaccard index.

Data were standardized in two ways to compare diversity trends within and between stands of both forests. First, because estimates of diversity are sensitive to the area or number of plots sampled, means were compared using the same individuals at plot-level values of richness and heterogeneity. Second, to control for natural differences in diversity between two lowland forests, the mean of less-sensitive richness indices were compared.

Species richness generally relates to the number of species appearing within a specific area, used to assess species abundance in a clearly defined area of interest (Magurran, 1988). In contrast, the term of species richness is also, somehow, intended for a borderless area; i.e., an area size ranged from infinity to a small-sampled area. The latter definition is possible because species vary in the number of individuals per area (common to rare species) that contribute to spatial-related increasing richness. It is crucial when one scales up richness abundance from a sample-sized area to a site-specific or even to a landscape area. Therefore, most ecological studies express the former richness definition as *species density*, which is species incidence per area unit (Simpson, 1949), and then term the species counted per total individuals in an area unit as *species-individuals ratio* (Cannon *et al.*, 1998).

However, these two simple definitions face difficulty when these parameters are used to compare two or more communities where the samples are from populations with different individual densities. For example, in a natural forest, *species density* varies with species richness and mean density of individuals. Gotelli & Colwell (2001) demonstrated this pitfall by showing that the higher stem-density of a second-growth forest went along with greater species richness than the lower stem-density of an old-growth forest when their area-based species rarefaction curves were compared, but the result was reversed when the area-based rarefaction was re-scaled to an individual-based. Additionally, an invalid conclusion would arise when using the *species-individual ratio* to compare two or

57

more unequal individual-density populations. This is because the ratio assumes a linear correlation between richness and abundance that is true only for an extremely uneven population, where only one species is dominant (Gotelli & Graves, 1996). The main point regarding which approach is valid is the individual that conveys taxonomic information. Therefore, *species density,* which has an area basis, should be standardized to a standard number of individuals (abundance-based sample) for comparison purposes.

A simple species-area curve is mainly used to determine the characteristic total number of species in a specific plant community (which is achieved when further enlargement of the area does not lead to a further increase of species numbers (Cain, 1938; Lamprecht, 1989); thus, this curve is not developed for comparison purposes but for the determination of the so-called minimum area which is needed to include all characteristic species of a certain plant community. Moreover, the species accumulation curve is constructed by one particular random ordering of all samples. Therefore, the method's under-sampling effect inevitably biases the observed species count (Gotelli & Colwell, 2001), which leads to a higher deviation in presenting species richness and other diversity measures. In contrast, the species rarefaction curve represents the means of repeated re-sampling of all pooled individuals or abundance-based samples. It shows statistical expectations for the corresponding accumulation curve, which are rarefaction and extrapolation (Colwell *et al.*, 2012).

Therefore, the specific objective was to present the species richness of the studied stands based on abundance data, which was comparable to various stands locally and regionally (even at a global level) by high-accuracy statistical analysis.

Species abundance data from subplots were then re-sampled with 150 runs for sample order randomization (without replacement) until all reference samples were accumulated. The expected number of species below the reference sample area (interpolation), desired species richness above the reference sample area (extrapolation), and their variation at a 95% confidence interval (CI) were calculated using the equations 4.2, 4.3, 4.4, and 4.5, respectively, following Colwell *et al.* (2012). The iterations and all data analyses were executed in EstimateS 9 software (Colwell, 2013).

$$\bar{S}_{area}(a) = S_{obs} - \sum_{X_i>0}\left(1 - \frac{a}{A}\right)^{X_i} \qquad 4.2$$

$$\sigma^2_{area}(a) = \sum_{k=1}^{n}\left[1 - \left(1 - \frac{a}{A}\right)^{k}\right]^2 f_k - \bar{S}_{area}(a)^2/S_{est} \qquad 4.3$$

$$\bar{S}_{area}(A + a^*) = S_{obs} + \hat{f}_0\left[1 - exp\left(-\frac{a^*}{A}\frac{f_1}{\hat{f}_0}\right)\right] \qquad 4.4$$

$$V\hat{a}r\left(\hat{S}_{area}(A + a^*)\right) = \sum_{i=1}^{n}\sum_{j=1}^{n}\frac{\partial \bar{S}}{\partial f_i}\frac{\partial \bar{S}}{\partial f_j}c\hat{o}v(f_i, f_j) \qquad 4.5$$

$$c\hat{o}v(f_i, f_j) = f_i[1 - f_i/(S_{obs} + \hat{f}_0)] \quad for\ i = j$$

$$c\hat{o}v(f_i, f_j) = -f_i f_j/(S_{obs} + \hat{f}_0) \quad for\ i \neq j$$

Where:

$\bar{S}_{area}(a)$ = Expected number of species in a random area of size a within the reference area of size A based on species abundances x_i in the reference sample.

$\sigma^2_{area}(a)$ = An unconditional variance of rarefied richness $\bar{S}_{area}(a)$.

$\bar{S}_{area}(A + a^*)$ = Expected no of species in an augmented area $A + a^*$ $(a^* > 0)$.

$v\hat{a}r\left(\hat{S}_{area}(A + a^*)\right)$ = Variance estimator for $\bar{S}_{area}(A + a^*)$ (as $\bar{S}$ in the formula).

S_{obs} = Total number of species observed in all samples

S_{est} = An estimator of the full species richness of the assemblage

$\hat{f}_0$ = Estimator for f_0, a function of the frequencies $(f_1, f_2, \ldots\ldots,f_n)$

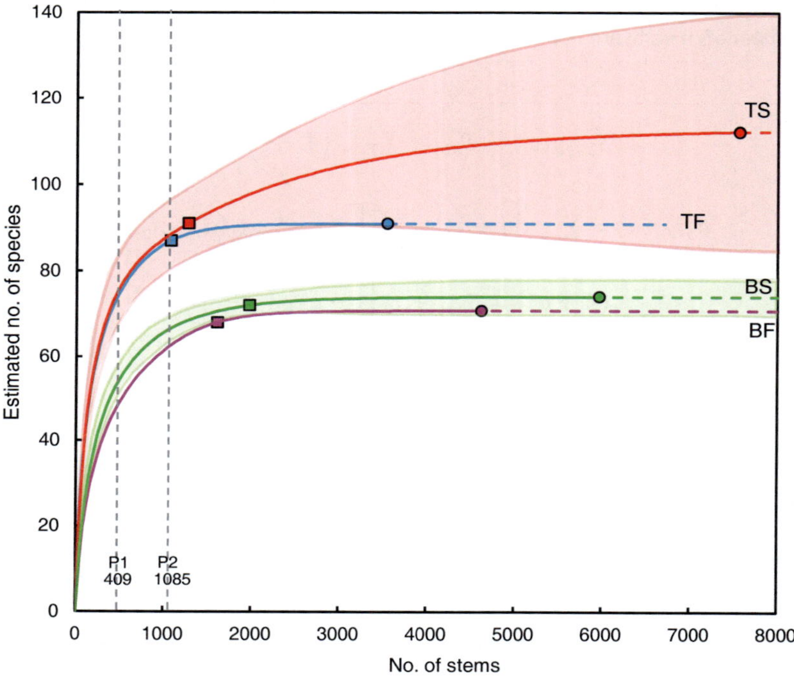

The solid line represents the rarefaction curve, which estimates species down from filled squares, and the extrapolation curve, which predicts species up from filled square to filled circle. Dash line curves show asymptote. Shaded areas are the confidence interval (95%) of the higher richness stand of each forest. Filled-squares are reference points represented by total species and individuals recorded in a 4-ha plot. Filled-circles reveal asymptote species richness and the corresponding individuals. P1=409 stems referred to the highest number of individuals counted at the 1-ha plot of BS stand. P2=1085 stems related to the lowest estimated individuals tallied at the 4-ha plot of TF stand. All tree species with DBH ≥ 10 cm were recorded from 40 0.1-ha subplots for each data set.

Figure 4.1. Species rarefaction and extrapolation curves for four study stands of Tunas and Bonggo.

Figure 4.1 shows that all curves rose rapidly up to P1 and P2 and continued to increase gradually until they flattened and reached the asymptotes (filled circles). This is usually the case that the species richness of stands continues to rise if the enumeration is carried on above the observed individuals at the plot level up to the maximum (asymptotic) richness at the stand level. The curves estimated that TS stand performed the highest richness of 112.17 (SD ± 14.06) species. It was followed by TF stand (90.93 ± 3.25 species), BS stand (73.95 ± 2.21 species), and BF stand (70.71 ± 2.46 species). However, the extrapolation curves levelled off soon after the reference point (filled square=max individuals recorded at plot level), meaning that the sampling method was sufficient to characterize the community. As a comparison, the observed richness found in this study, from TS, TF, BS, and BF stands, were 90, 87, 72, and 68 species, respectively. Statistically, the observed richness and the estimated richness were not remarkably different.

The graphs also depict that the species estimated curves of Tunas plots increased more rapidly and more linearly before reaching their reference points (filled squares) than did the estimated curves of the Bonggo plots. The sharply-increasing curve indicated that the plots were composed of a higher number of species per stems, while the linear curve showed that fewer species dominated the sampled areas. Indeed, the steeper and near-linear curves showed fewer species with many individuals (highly abundant species) and more species with fewer stems occupying the Tunas stands than the Bonggo stands. Furthermore, the CI of the TS stand curve was becoming significantly broader when rarefying at a higher number of individuals. This uncertainty in predicting additional species above the reference point indicated that a highly random number of species would be added or reduced if the richness was upscaled above the plot level due to the presence of many rare species.

A mean-test of species richness between two plots (flat and steep areas) was conducted at the same number of individuals for rarefaction means. P1 or 400 individuals was used as an equivalency to average stem densities at 1-ha, and P2 or 1085 individuals was another point of comparison referring to the reference point of the least-individual stand (TF). When rarefying from infinity to P2, the TF stand's richness (88.13±2.0 species) was similar to that of the TS stand (89.86±4.16 species). At the same individual level, the BF and BS stands showed a significant difference in richness (64.39±1.68 species and 67.64±1.62 species, respectively). Further tests indicated that after rarefying to P1, the tree richness of TF and TS stands remained similar (71.07±2.58 and 72.84±4.09 species, respectively). Meanwhile, BF stands (45.30±2.52 species) invariably showed significantly lower tree richness compared to that in BS stand (50.36±2.41 species).

In summary, the tree richness of the Tunas forest was higher than that of the Bonggo forest. Even though there was uncertainty in extrapolating additional species above the reference point, the study might assume that the TS and TF stands would have similar tree richness at the stand or landscape level due to the presence of many rare species. On the other hand, the BS and BF stands would show a significant richness difference between stand and landscape levels due to low species individual ratio.

4.2.2. Species diversity indices

This study provides diversity measures of four stands based on two different abundances, individuals and basal areas (BA), rarefied at P2 (1085 stems and 85.81 m^2), as shown in Table 4.6. The results reveal that the diversity indices performed various heterogeneity comparisons between stand communities because of their different sensitivity to interpret species abundance distributions.

At both, individuals- and BA-based abundances, the log series α indices were similar to those of species richness S in significantly discriminating tree diversity between Tunas and Bonggo stands. The α and S also had more similarities in giving no significant difference between stands within every forest. An α value, which is approximately equal to the number of species with a single individual in an assemblage, more likely relates to higher rareness in the stand with more species (Tunas stands) than that of fewer species (Bonggo stands), as explained by the richness S of the respective stands. In a particular case, where the species richness S was significantly different, the α index could not differentiate the individuals-based diversity because both stands had a relatively similar number of rare species, but not common ones. The richness of the flat and steep stand of Bonggo was an example.

Among the Shanon indices (H', E, $e^{H'}$, and $e^{H'}/S$), only the Shannon exponential $e^{H'}$ showed significantly higher diversity values of the Tunas stands than those of the Bonggo stands, based on both individuals- and BA-abundances. Shannon $e^{H'}$ is equivalent to the absolute number of equally common species required to produce Shannon H', as well as the Shannon differential indices of E and $e^{H'}/S$. These indices mainly show the proportion of equally common species, which are richness-dependent indices. Therefore, $e^{H'}$ is more sensitive in separating diversities of stand communities with various species. The finding indicates that the $e^{H'}$ discriminated not only diversities between stands with unequal species richness (between Tunas and Bonggo) but also differentiated the stands with relatively the same number of species (between flat and steep areas).

Table 4.6.   Diversity indices and their comparison between and within forests of Tunas and Bonggo in West Papua Indonesia

Diversity indices (n-based)	t_{forest}	Tunas Flat (mean± SD)	Tunas Steep (mean± SD)	t_{slope}	Bonggo Flat (mean± SD)	Bonggo Steep (mean± SD)	t_{slope}
Individuals		1085	1085		1085	1085	
S	10.49**	87.00±1.940	88.67±4.130	0.78[ns]	62.55±1.750	66.76±1.690	2.26*
α	4.41**	22.46±1.340	22.84±1.340	0.23[ns]	14.44±0.930	15.71±1.000	0.91[ns]
H'	0.63[ns]	3.86±0.010	3.91±0.020	0.29[ns]	3.51±0.040	3.72±0.040	0.74[ns]
E	0.25[ns]	0.86±0.002	0.87±0.004	0.09[ns]	0.85±0.004	0.88±0.004	0.41[ns]
eH'	4.25**	47.54±0.330	49.99±0.820	2.29*	33.50±1.350	41.32±1.810	4.39**
$(eH')/S$	0.19[ns]	0.55±0.008	0.56±0.016	0.10[ns]	0.54±0.006	0.62±0.011	0.63[ns]
1/D	0.55[ns]	26.60±0.480	29.83±0.930	2.72**	22.88±1.170	30.74±1.820	4.55**
(1/D)/S	0.26[ns]	0.31±0.001	0.34±0.005	0.38[ns]	0.37±0.008	0.46±0.015	0.62[ns]
(BA-based)							
BA (m2)		85.81	85.81		85.81	85.81	
S	9.28**	87.00±1.940	89.08±4.140	0.84[ns]	67.15±1.610	69.83±1.480	1.52[ns]
α	3.26**	15.94±0.740	16.33±0.750	0.32[ns]	11.60±0.590	12.15±0.610	0.50[ns]
H'	0.98[ns]	3.90±0.010	3.98±0.010	0.57[ns]	3.63±0.010	3.73±0.020	0.57[ns]
E	0.24[ns]	0.87±0.000	0.89±0.010	0.14[ns]	0.86±0.002	0.88±0.000	0.29[ns]
eH'	7.49**	49.19±0.000	53.48±0.720	5.06**	37.71±0.540	41.78±0.980	3.30**
$(eH')/S$	0.24[ns]	0.57±0.012	0.60±0.018	0.20[ns]	0.56±0.005	0.60±0.001	0.45[ns]
1/D	4.22**	31.30±0.340	35.88±0.760	4.37**	27.34±0.540	29.39±1.110	1.59**
(1/D)/S	0.59[ns]	0.36±0.004	0.40±0.009	0.37[ns]	0.41±0.002	0.42±0.007	0.15[ns]

S = Number of species, α = Log series index, H' = Shannon diversity, E = Shannon evenness, $e^{H'}$ = Shannon exponential, $(e^{H'})/S$ = Shannon exponential evenness, 1/D = Simpson invers diversity, (1/D)/S = Simpson inverse evenness. T_{forest} = t-test for mean between forests; t_{slope} = t-test for mean between slopes within forest; Mean difference at t-test > 0.05 = no significant (ns), < 0.05 = significant (*) and < 0.01 = highly significant (**): All the mean and standard deviation were resulted from 100 runs of rarefaction without replacement using EstimateS (Colwell, 2013).

The Simpson index 1/D was most likely equal to the Shannon $e^{H'}$ in its ability to explain diversity differences between the four stand communities. Among others, the richness-independent indices showed significant differences between tree communities of flat and steep areas that had a relatively similar number of species. The result indicated that even though the two tree communities had the same tree species richness, the species abundance distribution of the assemblages can be different. This is because the two indices weigh to abundances of the most common species and usually refer to dominance or evenness, unlike others that emphasize the species richness component of diversity. It was evident that when the measures weighed back to the richness (see their reciprocal indices of $e^{H'}/S$ and $(1/D)/S$), the result showed no significant differences in diversity between all stands.

However, the $e^{H'}$ and 1/D were different in ranking the diversity of the four tree communities. The $e^{H'}$ placed the diversity of TS stand in the top rank, followed by TF, BS, and BF stand; the rank of $e^{H'}$ values were consistent in both bases, individuals- and Basal-Areas. 1/D not only showed a different sequence of diversity rank for the four communities but also ranked the diversity of the four stand communities differently, according to their abundance and BA bases. Individual-based 1/D put the BS-stand in the first position, the TS- and TF-stand at the next lower places, respectively, and the BF-stand at the lowest. In contrast, BA-based 1/D showed the sequence of diversity values of the four tree communities to be similar to the rank order revealed by the $e^{H'}$. It resulted from markedly increased diversity values from individual to BA bases in the TF, TS, and BF, but that of the BS stand remained the same for both abundance bases.

4.2.3. Species similarity

The study presents species similarity between Tunas and Bonggo stands in Jaccard's Indices (Table 4.7).

Table 4.7. Species similarity between stands of Tunas and Bonggo forests presented in Jaccard-classic, Jaccard-individuals, and Jaccard-Basal Area calculated from tree species with DBH ≥ 10 cm.

Compared stands	J-Classic	J-Individuals		J-Basal area	
		Estimate	Residual	Estimate	Residual
TF – TS	0.74	0.96	2.0	0.94	0.2
TF – BF	0.25	0.37	0.6	0.32	0.0
TF – BS	0.25	0.37	0.1	0.34	0.0
TS – BF	0.23	0.38	2.2	0.33	0.0
TS – BS	0.25	0.38	1.2	0.34	0.0
BF – BS	0.82	0.98	1.5	0.94	0.1

TF = Tunas flat; TS = Tunas steep; BF = Bonggo flat; BS = Bonggo steep. J-Classic was Jaccard index calculated from binary data (species' presence-absence); J-Individuals and J-Basal area were adjusted Jaccard indices derived respectively from the estimated relative abundance of individuals and basal-areas (Chao *et al.*, 2000).

The table clearly shows that species similarity indices for stands within the same forest (TF-TS or BF-BS) were markedly higher than those for comparing stands between different locations (TF-BF, TF-BS, TS-BF, or TS-BS). At the plot level, the TF- and TS-stand of Tunas forest shared a high number of species observed (J-classic of 0.74). This was also the case for the BF- and BS-stand in the Bonggo forest (J-classic of 0.82). More importantly, adjusted Jaccard's indices revealed remarkably higher similarities to those comparable stands. The increase of adjusted values implied that the abundance distribution of every species observed in the plots plays a vital role in estimating the stand-level species unobserved at plot-level measurement, i.e., they were mostly species with a single individual.

4.2.4. Tree diversity in comparisons to other sites in New Guinea

The tree species richness in the two study forests is within the range of tree species numbers in the lowland forests of New Guinea. The observed tree species at plot level were 90 in Tunas plots and 72 in Bonggo, while the estimated species for both stands were about 112 and 74 tree species in Tunas and Bonggo forests, respectively. Marwa (2009) investigated 68 tree species within 1449 stems with DBH>10cm in the primary forest of the Tor-Apauwer logging concession, Sarmi. This humic-ferrasols forest is situated at steeply dissected to mountainous with a slope of 20-30% in the northern lowland of the Indonesian part of New Guinea. It is the lowest tree species richness area ever reported from the lowland forest of Western New Guinea. The highest tree species richness among lowland forests in the region is in the logging concession of Sawaerma primary forest in the southern lowland of Indonesian Papua. Kuswandi (2017) encountered 121 species from 1205 trees with DBH>10cm in the Sawaerma primary forest, which is dominated by ferric-acrisols (reddish-yellow podzolic) soil in a flat to an undulating area with a slope of 8-15%. Other lowland forests of the west New Guinea, such as the Bird's Head area of New Guinea, had a range of tree species richness for some primary forests in logging concession areas between those in the mainland (Sarmi and Sawerma). For example, a study reported 70 species found in the district-fluvisols (alluvial-soil) forest with mainly red-yellow sand at the logging site of (Prafi) Manokwari, northeastern Bird's Neck area (Kuswandi *et al.,* 1993). Another study investigated 109 species from orthic-acrisols (red-yellow podzolic) soil forest with a mostly smooth texture fraction at Sungguan Valey of the northwestern part of the region (Wagner, 2016).

The range of tree species richness in Eastern New Guinea is also comparable with those depicted above. Johns (1985) described 55 tree species in a large timber harvesting block in Gogol - Madang, in the lowland forest of the northern Central Ranges of Papua New Guinea (PNG). The low number of tree species was because only stems with DBH > 20cm were measured. Kiapranis (1991) reported a significantly higher richness of 180 tree species in 6,269 stems from the same Gogol area by including woody seedlings, but did not provide specific information for individuals with DBH>10cm. Eutric-fluvisol is the dominant soils in the Gogol valley - the Madang Province. The soil texture varies from sandy-loam around the flat area (near river bench) to silty-clay and clay near the foothills (FAO-Unesco, 1974). Even though there was limited data available regarding tree species from the southern part of the PNG, Oatham & Beehler (1995) did study trees and lianas from a priority area for biodiversity conservation in the Lakekamu Basin - Southern Papua New Guinea. The study provided a combination of tree and liana data of 149 species in 426 stems with DBH>10cm recorded in the Nagore-north plot, 178 species of 482 trees in the Nagore-south plot, and 93 species of 392 trees in the Sii plot.

Many factors drive tree species richness variations, and the findings of this study and its comparison with other studies mentioned previously imply that the discrepancy of tree species richness in the moist lowland forests of New Guinea is most likely due to soil drainage and nutrition. The lower richness forests such as in Bonggo, Tor-Apauwer, Prafi, and Gogol grow on coarse- and middle-textured and dominant district-cambisols, humic-ferrasols, district-fluvisol, and eutric-fluvisols, respectively, which are generally high in organic matter and nutritions (FAO-Unesco, 1974). The higher tree species richness of lowland forests of Tunas, Sawaerma, and Sungguan have fine-textured, nutrient-limited soils such as ferric-acrisols and orthic-acrisols (FAO-Unesco, 1974). However, Oatham & Beehler (1995) investigated higher woody stems richness in Lakekamu Basin plots where the world soil map (FAO-Unesco, 1974) identifies middle-textured and district-cambisols soil dominating the area. Although they included lianas in the measuring data, the species richness of the Lakekamu remains exceptionally high compared to the tree species richness from the plots with identical soil texture and nutrition (e.g., Bonggo plots).

The Lakekamu basin is an enclaved area of 500 km^2 surrounded by 2000 km^2 of hill and montane forest on a complex of lowland alluvial plain, fan, and swamp formed by the Lakekamu River (Oatham & Beehler, 1995). In the basin, the Nagore plots, which are located on stony outwash alluvium, represented higher woody species (149-178) than that (93) of the Sii plot, which was fine silt alluvium

(Oatham & Beehler, 1995).   They claimed higher diversity in the basin is closely related to canopy species' dispersion rather than sub-canopy species, which are more clustered. According to Paijmans (1976), prolonged inundation during wet-season in the lowland alluvial plain of New Guinea is mainly associated with a lower and smaller crowned forest canopy with more large gaps, more widely space emergent, and a lower number of trees with small DBH than in forest on well-drained terrain. It is a favourable habitat for hydro-seral species and thick lianas with adventitious roots (in response to inundation) that increase the woody stems richness.

From the information provided above, some major important factors are presumed to cause the high variation of tree species diversity in New Guinea, such as the high species alteration over geographical distance (beta-diversity), the different geological history (coastal, alluvial plain vs. inland, raised land) resulting in habitat differences (drainage, soil nutrient status), and the local vegetation history (natural and/or anthropogenic disturbance).

4.3. Stand structure and dynamics in the primary forest

4.3.1. Stem density, mean diameter, and basal area (BA)

The tropical rain forest is exceptionally complex in structure (Richards, 1996). Its structure consists typically of three to four horizontal storeys of woody tree species and two storeys of shrubs, herbaceous plants, and a regeneration layer of woody tree species (Lamprecht, 1989). The uniqueness of storeys can help understand forest ecosystems' status (history, function, and succession) since each storey develops through ecosystem processes that vary among forest communities (Spies, 1998).

The mean diameter and height of trees in a stand are the main parameters used in assessing a forest. Nevertheless, the stand's average tree diameter and height are static parameters with limited information about a forest's status and its management purposes. Therefore, van Laar & Akça (2007) suggested more dynamic parameters such as quadratic diameter, central-basal-area diameter, and Lorey's mean height, which are meaningful in defining forest statuses.

The diameter-class distribution of tree species is another component of the stand structure that is very important in describing forest characters and can contribute to a better knowledge of the structural patterns in a tropical rainforest (Huang *et al.*, 2003). This parameter is the most useful tool for managing an uneven-aged stand (Isango, 2007) because it closely relates to many other structural features and habitat elements (Spies, 1998).

The tree diameter-class distributions of tropical forest stands vary considerably and depend mainly on environmental gradients, such as light traits (e.g., light-demanding, shade tolerant, and semi-shade tolerant), levels of natural disturbance, and human interferences (Poorter *et al.*, 2008). A curve of population structure derived from the tree size distribution is an appropriate variable to explain the current process of stand development. Accordingly, the shape of the population curve can help interpret tree species' history and future in a forest stand (Richards, 1996; McCune *et al.*, 2002; Kimmins, 2004; Poorter & Bongers,

2006; Hung, 2008). Therefore, this section provides the stand structure characteristics of the four study stands, including horizontal and vertical species distributions. The findings are valuable information with silvicultural implications for planning processes such as harvesting or thinning and selecting tree species for enrichment planting or rehabilitation purposes.

The mean stem density, diameter, and basal area (BA) of trees recorded in four stands of Tunas and Bonggo forests are depicted in Table 4.8.

Table 4.8.   Stem density, mean diameter, and BA of all trees with DBH $\geq$ 10 cm in four stands of Tunas and Bonggo forests.

Structural characteristics	Stands				F values
	TF Mean±SE	TS Mean±SE	BF Mean±SE	BS Mean±SE	
No.of stems (n ha^{-1})	268.50^a±9.7	319.3^b±4.9	402.0^c±18.3	493.3^d±7.6	76.05[***]
Diameter (cm):					
- *Arithmetic*	28.43^a±0.4	28.11^a±0.2	24.54^b±0.3	25.34^b±0.3	47.77[***]
- *Quadratic*	31.93^a±0.5	31.83^a±0.1	29.39^b±0.4	29.97^b±0.3	14.47[***]
- *20% largest trees*	51.34 ±1.1	53.27 ±0.9	47.10 ±5.5	53.70 ±0.7	1.34[ns]
- *Central basal area*	45.55^a±0.6	47.41^a±0.9	48.05^a±0.9	51.47^b±0.7	10.47[**]
BA (m^2 ha^{-1})	21.45^a±0.4	25.40^b±0.6	27.21^b±0.7	34.55^c±0.4	112.7[***]

Mean values and standard error (SE) of the number of stems, mean tree diameter of the stand (arithmetic diameter, quadratic diameter, the diameter of 20% largest trees and diameter of central basal area tree), and basal area were calculated from four 1-ha subplots for each of the four stands. The differences among plots were analyzed using ANOVA for grouping variables (STANDS) indicated by F values, and significant (p) of the tests are provided. * p$\leq$0.05; ** p$\leq$0.01; *** p$\leq$0.001; ns= non-significant. Post-tests (Tukey HSD) for the mean are indicated by a, b, c, d in mean value.TF=Tunas flat; TS=Tunas steep; BF=Bonggo flat; BS=Bonggo steep.

As shown in Table 4.8, the tree population densities were higher in Bonggo stands than in Tunas, but the denser tree stems were consistently found in both forests' steep stands. Accordingly, the most populated tree community was BS stand, with 493 stems ha^{-1}, and the lowest was TF stand, with 268 stems ha^{-1}. The densities of Tunas stands are slightly below the range of tree densities currently reported for New Guinean primary lowland forests, i.e., between 392 stems ha^{-1} and 560 stems ha^{-1} (Paijmans, 1970; Johns, 1985; Kiapranis, 1991;

Oatham & Beehler, 1995). However, the findings are within the range of the world's stem densities for primary moist lowland forests, which vary significantly from 245 ha^{-1} to 720 ha^{-1} for trees with DBH>10cm (Richards, 1996).

In line with tree density, the most populated stand BS had the highest BA of 34.55 m^{2}ha^{-1} and the least dense stand TF was associated with the lowest BA of 21.45 m^{2}ha^{-1}. Nevertheless, the increase of the stem BA with tree density was not always linear. The findings indicated that the stand BF was substantially different than the stand TS in tree density, but their BAs were statistically similar. This figure implies that variation in tree size composition among stands is essential to explain the density-BA correlation.

The study found that arithmetic diameter means were markedly lower in Bonggo stands than in Tunas stands. This is because the Bonggo stands were composed of a higher abundance of small-sized trees. The presence of many small trees in the Bonggo stands was more noticeable from the values of quadratic means, i.e., a calculation that weights mean diameter toward large-sized trees. The stand's mean values increased more significantly from arithmetic to quadratic means, by about 5 cm, compared to a 3 cm increase in Tunas stands. Another finding in tree size composition was that the mean diameter of the 20%-largest trees was not significantly different among the four stands. However, this mean diameter is a density-depended value with a high standard deviation if a stand has a high-density population with a broader range of diameter classes.

Further critical stand's mean diameter is the mean diameter of central-basal-area (CBA), which explains which tree diameter size contributes to the highest portion of the basal area in a stand. The result showed that the BS stand showed the most top mean diameter CBA of 51.47 cm. It was far from the values of BF (48.05 cm), TS (47.41 cm), and TF stands (45.55 cm). This finding showed that trees with DBH larger than 50 cm contributed to the outstandingly high BA of the BS stand.

4.3.2. Diameter distribution and slope of population structure

Tree abundance distribution based on diameter classes is a crucial parameter in describing forest structure (Loetsch *et al.*, 1973). It is also useful in developing a stems-diameter table for estimating stand volume (van Laar & Akça, 2007). More importantly, the slope of the population structure and the length of tail of the tree diameter-class distribution curve are the best approaches to analyzing the dynamic processes of a stand (Weidelt, 1998). Accordingly, this study determined the slopes of population structure and the number of classes with tree DBH bigger than 50 cm from the four reviewed stands: Tunas flat (TF), Tunas steep (TS), Bonggo flat (BF), and Bonggo steep (BS).

Among various mathematical models, an inverse J-shape curve of diameter distribution is most favoured for mixed natural forests (Meyer, 1953; Philip, 1994). The number of trees in each diameter class (y-axis) was plotted against the central numbers of the respective class interval of 10 cm (x-axes), as indicated by the bar graphs in Figure 4.2. An exponential equation $N = a \cdot e^{-b \cdot x}$ was then applied to fit the observed tree distribution. N is the number of trees in a diameter class, e is Euler's number (2.71828), a is a constant number limiting the maximum number of trees at the smallest diameter above DBH>0, and b is a constant number explaining the slope of the curve. A matrix correlation was used to test the fitness of observed and estimated values at $p = 0.05$ and a confidence interval (CI) of 95%. The result indicated that the estimated values of all stands fitted the observed values well, with a significant relationship $p < 0.05$ at CI 95%; the coefficients of correlation (R) of all stands were above 0.986, indicating that the model can effectively explain (98%) the observed stems for every class.

The patterns of tree size distributions for the four studied stands are depicted in Figure 4.2. The constants a and b in the negative exponential equations and the numbers of classes above 50 cm were varied, indicating a distinct distribution of tree sizes between the stands.

The graphs show that the tree size distribution of all stands followed a familiar pattern of a typically inverted J-shape curve for a mixed forest; the number of individuals is highly abundant for the smallest diameter and decreases as the size increases. With copious stems concentrated at the small-sized classes, the Bonggo stands showed a higher percentage of trees with DBHs of less than 20 cm compared to that in Tunas stands, specifically 47.51% (BF stand), 43.59% (BS stand), 36.22% (TF stand) and 36.50% (TS stand). For the next-higher class (DBH 20-30 cm), the number of stems also dropped more sharply in Bonggo stands than in Tunas stands. Additionally, within forests, the BF stand indicated a steeper reduction of tree population from the smallest to the next larger classes than the BS stand, which was confirmed by the higher negative slope (b-value) structure, respectively -0.065 -0.055. On the other hand, the two stands of the Tunas forest represented a similar gradually sloping population structure (-0.044).

Moreover, all stands revealed that the large size classes (>50 cm DBH) represented less than 9% of the total tree abundance with a rather similar distribution of DBH classes up to 110-120 cm. Due to considerable differences concerning the lowest DBH classes, the Bonggo stands represented a significantly lower percentage of trees with DBH>50 cm (6.90% in flat and 6.54% in steep areas) than those in Tunas stands, represented by 8.66% 8.60% of total trees. Besides being higher in absolute numbers (28 and 32 stems), the large trees (DBH>50cm) of Bonggo stands were distributed among more diameter-size classes (7 and 8) than those of Tunas stands, which had 23 and 28 stems spread among only 4 and 5 diameter-size classes. The biggest trees in Bonggo reached diameters of 125.0 cm (in the BS stand) and 119.0 cm (in the BF stand), and were represented by the same species, *Pterocarpus indicus*. In comparison, those of Tunas reached diameters of 96.2 cm (*Pometia pinata*) and 86.9 cm (*Dacriodes sp*) in BS and BF stands, respectively.

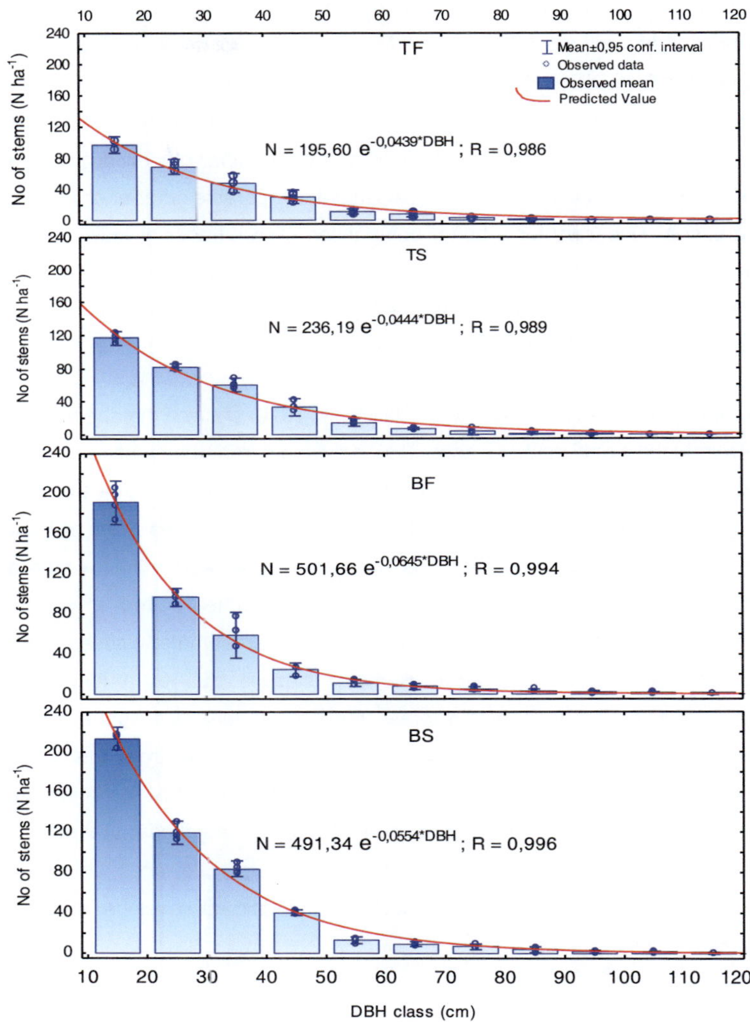

Mean value and 95% CI of no.stems (Nha-1) in every DBH class (observed data) in stands TF, TS, BF, and BS indicated in bar graph were calculated from all trees with DBH>10cm in four 1-ha subplots (n=4). A reverse J-shape model, N=ae-bx, was applied to estimate diameter distribution function using Nonlinear Regression Analysis. Correlation of observed and predicted values (R) is provided together with the estimated functions.

Figure 4.2. Observed and predicted stems in each 10-cm DBH class.

Rollet (1974) found that in the natural rainforest of Venezuelan Guiana, the distribution of trees with girth-classes fits an exponential model well, but only for stems of $\geq$20cm DBH. His graph also insists on a sharp decrease, although the slope slowly flattens at larger classes. However, he found that the model's fitness is less when tree size-class <20cm DBH and the number of individuals from an area >1 ha are predicted. This suggests that a comparable number of individuals or areas is necessary for a better comparison. The number of individual trees used in this diameter distribution analysis widely varied between Tunas and Bonggo forests; nevertheless, the same size of the area sampled is more than necessary for this stand comparison.

In four 0.8-ha plots of New Guinea, (Paijmans, 1970) showed that the number of trees reduced significantly by about 260 individuals from the smallest class size (>20 cm DBH) to the next larger class (20-30cm DBH), and only three stems were recorded in size group > 64.7 cm DBH. Tree abundance distributions of Bonggo and Tunas stands are within the range of Paijman's plots. According to Lamprecht (1989), mature forests follow of tree diameter distribution pattern in which the stem number drops sharply from small to next higher diameter classes. Then, it gradually reduces to very large diameter classes due to a long natural regeneration period in the virgin forest. Richards (1996) also agrees with this typical pattern. Still, he reports considerable divergences from this trend in tropical forests because of different histories and seral status, including highly diverse tree species. This is in agreement with the observation that the Bonggo and Tunas forests have developed from two geological provinces with different soil formations, which led to the vegetation's specific successional status and diversity.

4.3.3. Species distribution in tree size classes

Growth and mortality rates at each stage of tree development clearly explain the various patterns of stem width distribution of moist forest; however, variation between individual trees and species remains little known (Alder, 1995; Richards, 1996). Therefore, the study presents numbers of trees and species that successfully survived and entered the early (DBH 10-30 cm), middle (30-50 cm), and last stages (>50 cm) (Table 4.9).

Table 4.9. The percentage and number of stems and species grouped in five size classes of the four study stands

Stands	Diameter classes (cm)									
	10-30	30-50	50-70	70-90	>90	10-30	30-50	50-70	70-90	>90
	Stems ha^{-1}					*Species ha^{-1}*				
TF	62.01	29.33	6.98	1.68	-	93.10	72.41	17.24	5.76	-
	167	*79*	*19*	*5*	*-*	*81*	*63*	*15*	*5*	*-*
TS	62.26	29.13	6.73	1.72	0.16	92.31	78.02	16.48	5.49	1.01
	199	*93*	*22*	*6*	*1*	*84*	*71*	*15*	*5*	*1*
BF	71.58	20.77	4.66	2.24	0.75	82.35	80.88	25.00	11.76	4.41
	288	*84*	*19*	*9*	*3*	*56*	*55*	*17*	*8*	*3*
BS	67.75	24.95	4.26	2.03	1.01	88.89	79.17	23.61	12.50	5.56
	334	*123*	*21*	*10*	*5*	*64*	*57*	*17*	*9*	*4*

Percentage values printed in regular characters; Absolute values written in italic characters. TF=Tunas flat; TS= Tunas steep; BF=Bonggo flat, and BS=Bonggo steep.

It is clear from Table 4.9 that most of the tree species were concentrated in the establishment phase of tree growth (DBH 10-30 cm). The species proportion in the first-stage ranged from 82% (BF stand) to 93% (TF stand). In general, around 90% of tropical tree species recorded in a sample plot occupy the smallest diameter class (Cam. 2015). On the contrary, less than 6% of all species could eventually reach the mature phase (DBH >70 cm in Tunas forest and > 90 cm in Bonggo forest).

From the establishment to the early building phase (DBH 30-50 cm), the gradual reduction in all stands' species was not proportional to the decrease in the stem number. The BF stand, with a sharp decline in individuals, decreased by one species (2%), in comparison to the TS stand, which had a gradual drop of stems that showed a significant decrease by 18 species (21%) that unsuccessfully began growth at the building stage. This shows that more of the establishing species in Bonggo stands had the opportunity to grow in the building phase than those in Tunas. Also, common species (with highly abundant individuals) had more changes than the rare species; Bonggo stands (with denser stems and lower species number) were composed of many common species at the establishment stage.

The tree species of both forests dropped suddenly to around 17% (in Tunas) and 24% (in Bonggo) in the late building phase (DBH 50-70 cm). Trees at this size class must compete with each other to put their canopy at a height that can receive direct light. Species that love shade remained slow growing and even died under the crown of others. Alder (1995) stated that a sharp decline of distribution is due to high mortality and slow growth rates of individuals from shade-tolerant species with smaller DBH classes in tropical natural mixed forest.

On the other hand, species that need light for their growth kept growing and successfully entered the mature phase (DBH >70 cm). The study recorded approximately 14 species in the Tunas forest and 18 species in the Bonggo forest. Still, on average, there were only five and nine species per hectare, respectively, found in the two forests. (Oatham & Beehler, 1995) reported the dispersion of canopy species is more clustered than sub-canopy species. Further growth of these species mainly depends on the physiological characteristics (long lifespan), the forest's age, and soil characteristics (Richards, 1996). The Tunas forest's biggest tree reached a DBH of about 90 cm, and that of Bonggo showed the widest diameter of 125 cm. These long-life species of Bonggo could grow over 90 cm in DBH because the soils are nutrient-rich and well-drained (see chapter 2).

4.3.4. Stand height

The mean heights of trees investigated in four 1-ha plots for each of the four stands in Tunas and Bonggo forests are shown in Table 4.10. The mean height (arithmetic) of the study stands was significantly higher in Bonggo than in Tunas. The highest was the BS stand (17.34 m), and the lowest was the TF stand (15.08 m). There was no marked difference in the mean tree height of the Tunas forest's flat and steep areas. In contrast, those in Bonggo were significantly different, with the mean tree height in the flat area (16.17 m) about one-meter below that in the steep site.

The tallest trees' mean height in the stands indicated that the BS stand had the highest top-canopy trees. On average, it was approximately 31.97 m (100 tallest trees) and 31.52 m (20% of tallest trees). These canopy heights were outstanding since they were 8 and 7 m higher than the next taller stand of BF, and 12 and 11 m different from those of shorter stands in Tunas.

Table 4.10. Stand heights of all trees with DBH $\geq$ 10 cm in four stands of Tunas and Bonggo forests

Stand characteristics	Stands				F values
	TF $Mean \pm SE$	TS $Mean \pm SE$	BF $Mean \pm SE$	BS $Mean \pm SE$	
Mean height (m) :					
- Arithmetic	$15.08^a \pm 0.1$	$15.62^{ab} \pm 0.2$	$16.17^b \pm 0.3$	$17.34^c \pm 0.3$	35.13^{***}
- 100 tallest trees	$20.31^a \pm 0.6$	$21.79^{ab} \pm 0.3$	$23.68^b \pm 0.4$	$31.97^c \pm 1.1$	58.70^{***}
- 20% tallest trees	$21.85^a \pm 0.4$	$22.78^a \pm 0.5$	$24.05^a \pm 0.5$	$31.52^b \pm 1.1$	42.84^{***}
- Lorey's mean	$18.69^a \pm 0.2$	$20.09^b \pm 0.1$	$23.57^c \pm 0.4$	$29.13^d \pm 0.5$	205.22^{***}

Mean values and standard error (SE) for mean tree heights of the stands (arithmetic height, 100 tallest trees, 20% tallest trees, and Lorey's mean height) were calculated from four 1-ha subplots for each of the four stands. The differences among plots were analyzed using ANOVA for grouping variables (STANDS) indicated by F values, and significant (p) of the tests are provided. * $p \leq 0.05$, ** $p \leq 0.01$, *** $p \leq 0.001$, ns= non-significant. Post-tests (Tukey HSD) for means are indicated by characters a, b, c, d in mean value. TF=Tunas flat; TS=Tunas steep; BF=Bonggo flat; BS=Bonggo steep.

Apart from the arithmetic and the tallest trees means, Lorey's mean height showed that the BS stand had the tallest mean (29.13 m) of trees that contributed to the largest portion of BA in the respective stands. Even though located in the same forest, the BF stand had a Lorey's mean height about 7 m lower than that of the BS stand. Those of the Tunas stands were even lower by the differences of 9 and 11 m.

The stand heights recorded in this study are not remarkably different from others in New Guinea. Oatham & Beehler (1995) reported stand heights between 14 m and 19 m from three plots of Lakekamu Basin (one at Nagore and two at Sii River), Papua New Guinea. Wright *et al.* (1997) measured a mean height of 19 m in a 1-ha hill plot of Crater Mountain, Papua New Guinea. No other information was given from these two studies, but Oatham & Beehler (1995) stated that relatively short-stature trees dominated the Nagore plots and that the emergent trees were sparsely distributed in all the plots. This height of tree size composition indicated that stand height in the moist tropical lowland forests of New Guinea could vary significantly among stands.

4.3.5. Vertical structure distribution

Sunlight illumination (or radiation) and air circulation inside a forest canopy are vital for tree species' growth, regeneration, and mortality (Richards, 1996). Tree species respond differently to light intensity levels resulting from transmission and reflection by leaves, trunks, and branches under the canopy. As a result, tree species variously appear within the spectrum from shade-loving to light-demanding species, depending on the level of light intensities. Vertical distribution of tree density and species can serve as important information underlying dynamic processes including growth, regeneration, mortality, and understory development (Hao *et al.*, 2007). The vertical composition can also inform the timber harvesting planning of a production stand.

However, in a moist mixed forest, boundaries between imaginary layers (A, B, and C or more storeys) are generally vague (Paijmans, 1970; Lamprecht, 1989; Richards, 1996). This is most likely because of the high richness of the mixed rainforest (Paijmans, 1970). In some cases, total tree heights provide little information about stratification because the crown depth and its ratio to the forest's total height vary among individual trees and species (Ogawa, 1965; Paijmans, 1970). However, Klinge *et al.* (1975) indicate that there are different phytomass densities between arbitrarily defined layers. This statement agrees with a finding that the transition between layers is irregular, gradual, or even lacking distinction (Lamprecht, 1989). Consequently, this study presents stem and species densities of the lower, middle, and upper storeys of a 1-ha stand area and analyzes the vertical structure according to those of each forest layer's compositions in the four studied stands.

The vertical structure was defined following IUFRO's scheme (Lamprecht, 1989). A stand was divided into three storeys based on its maximum height. Trees with total-height belonging to the top one-third interval were grouped in the upper storey, followed by trees in the second one-third range in the middle storey and trees in the bottom one-third of stand height in the lower storey. The number of trees and species for each storey is shown in percentage of total population or richness to compare stands (Table 4.11).

Tunas and Bonggo stands differed in the number of stems counted in the stands' vertical layers. Most tree canopies (62.73% and 65.36%) were concentrated in the middle layer of the Tunas stands, while those of Bonggo stands (50.87% and 63.66%) populated the lower layer. This distinction of tree crown allocation in each storey, does possibly reflect differences in light exposure, i.e. in allowing light to penetrate the stand canopies.

Table 4.11. Relative and absolute abundance of individuals and species with DBH ≥10 cm distributed in three forest storeys in four study stands of Tunas and Bonggo forests.

| Stands | Canopy storeys | | | | | |
| | Lower | Middle | Upper | Lower | Middle | Upper |
	Stems ha^{-1}			*Species ha^{-1}*		
TF	12.66	65.36	21.97	32.18	91.95	64.37
	34	*176*	*59*	*28*	*80*	*56*
TS	12.45	62.73	24.82	34.07	92.31	73.63
	40	*200*	*79*	*31*	*84*	*67*
BF	50.87	46.58	2.55	72.06	89.71	11.76
	205	*187*	*10*	*49*	*61*	*8*
BS	63.66	32.74	3.60	87.50	83.33	16.67
	314	*162*	*18*	*63*	*60*	*12*

Percentage values printed in regular characters; Absolute values written in italic characters. TF=Tunas flat; TS= Tunas steep; BF=Bonggo flat, and BS=Bonggo steep.

Tunas flat and steep stands most likely transmitted more light through their crowns. This was because the stands had lower stem densities (268 and 319 trees ha^{-1}) and thinner canopy layers (each of 10 m), in comparison to the higher density stands (402 and 493 trees ha^{-1}) of Bonggo, with thicker canopy storeys (each of 20 m).   In a situation when the lower layer has more light available and less shading from neighbours, the tree trunk, as well as the crown, would grow in height and diameter to balance their mechanical standing.   Therefore, many trees in the lower storey of Tunas stands had more chances to reach the middle storey, where their crown prevented each other from getting to the higher level.

4.3.6. Diameter-height curves

The stand diameter-height curve is mainly used to predict tree height when only diameter is measured (Weidelt, 1998). In a tropical moist forest with tall, dense canopies and broad crowns, it is challenging to record the tree height (Larjavaara & Muller-Landau, 2013). Bias in measurement accuracy related to the equipment used, as well as the enumerator's skill, is another reason (van Laar & Akça, 2007). Stand diameter-height relationship is also used in evaluating the status of a forest stand, i.e., young vs. mature forest (Hung, 2008), site quality of three forests (Pham, 2012), and disturbed vs. undisturbed forest (Cam, 2015).

The stem diameter-height relation is generally determined by the curve of the diameter versus height plot, in which the pattern of allometric scaling exponent is explained at a specific time of measurement (O'Brien *et al.*, 1995). According to van Laar & Akça (2007), there are more than ten functions (i.e., quadratic, logarithmic, and exponential functions) that can fit this relationship, depending on the tree community's condition. This static diameter-height relationship is also widely used to interpret the diameter growth rate relative to the height growth rate of stems in a forest stand community. In general, the shape of the curve shows that at higher diameters (i.e., canopy trees), the diameter increases more rapidly than the height. On the other hand, at smaller diameters, the tree mostly grows preferentially to height.

It is questionable whether a given tree follows such an allometric relationship during its growth because a wide variation in growth rate is found among species and individuals of the same species and their stages in the development of most tropical rainforests (Whitmore, 1984). This high variability in growth rate contributes to more complexity when the mixed forest status is assessed, unless the growth and mortality of every species is fully understood (Richards, 1996).

Static data of individuals at one-time measurements have been applied to show if diameter-height allometry works in the natural tree population (Rai, 1979; Rich *et al.*, 1986). Most of these studies agree with biomechanical models, which scale diameter as the power of height greater than one (O'Brien *et al.*, 1995). However, only a few studies adequately considered the effects of shading and wind, which is firmly controlled by the surrounding trees (King, 1996). Moreover, (Niklas, 1995) indicated that the allometric scaling function varied with stand density in seven species. Also, most of these studies neglected the relative position of sampled trees in the canopy or under the effect of their neighbours.

A tree needs to increase its stem height to gain the advantages of more exposure of leaves to sunlight, avoidance of shading competitors, and more change of reproductive or dispersal organs at a higher level (Rich *et al.*, 1986). In maintaining its height, a stem must have enough strength to balance elastic deformation and avoid bending under its mass (Niklas, 1995). Consequently, the mechanical strength needs of the stem at a certain height should be previously supported by diameter growth to a minimum required size (Tilman, 2020). Nevertheless, in particular conditions, this biomechanical expectation disagrees with the neighbours' effect (Henry & Aarssen, 1999). A tree's estimated growth may be impaired by neighbours' appearance, causing a change in the irradiant level. In high-density assemblages, trees tend to focus growth energy on height rather than diameter to maximize their exposure to light, resulting in slender stems (Rich *et al.*, 1986; Oliver & Larson, 1990). Although neighbours reduce wind impacts that mechanically cause an increase in the diameter-height ratios, lateral shade has been shown to have a more substantial effect on increasing the size ratios in tree species (Holbrook & Putz, 1989).

Therefore, Henry & Aarssen (1999) proposed two arguments that may explain why the diameter-height relationship should be less steep in big-sized stems than in small-size trees. First, "biomechanical constraints" control the increase in the height of most canopy trees, which experience relatively less neighbour density, to maximize the diameter growth to above the minimum requirement for bending strength of the stems. Second, "light competition" controls the height increase of sub-canopy trees, which are primarily shaded by neighbours, causing them to grow faster in height than in diameter. They also emphasized that when growing, a tree does not express age and neighbour as separated effects, and that these effects are confounded. As can be illustrated in a suppression period, a late seral tree can wait for a longer time than an early seral tree before growing relatively in high or diameter to a certain size ratio after getting chances from gap formation. When a tree has enough physical space available in the canopy in the transition period, it concentrates mainly on crown spread and stem diameter growth, with relatively little further increase in height.

When static data is used to infer individual trees' dynamic growth trajectories in a stand, the curve is referred to age as an independent factor. However, in natural mixed forests, age and neighbours are expected to be confounded effects on tree growth. Hence, Henry & Aarssen (1999) suggested that the growth trajectory (of a stand) can be inferred from a static diameter-height relationship by applying two hypotheses. Firstly, the early-growth trees formed the canopy and followed the biomechanical model at the beginning of their life, since the neighbour effect was low. The next arrival trees gradually experienced more neighbour effects from their successors such that their diameter-height ratios increased. As a result, a static diameter-height relationship's shallow slope is shaped in a group of big (canopy) trees. Secondly, the growth of late group trees, which develop in sub-canopy, fit the light competition assumption, since the neighbour effect is high. They started their development under the shading of neighbours that also protected them from mechanical influences. After getting enough energy for growing from a series of gap formations, the successful trees spend their energy preferentially on height growth to maintain higher competence. These trees show a high diameter-height ratio. Upon approach of the canopy, the neighbour effect decreases and is confounded with the biomechanical effect, resulting in a rapid reduction of the diameter-height ratio. As a result, the slope of the static diameter-height curve in this area (small-sized diameter) is steeper.

Accordingly, in this study, the Chapman – Richards' growth function (Richards, 1996) was used to develop a static diameter-height relationship to explain the four dynamic growth stands in Tunas and Bonggo.

Data sets of all trees ≥ 10cm DBH in each of the four stands, Tunas flat (TF), Tunas steep (TS), Bonggo flat (BF), and Bonggo steep (BS), were plotted to show diameter (DBH) against height. A curvilinear relationship was chosen because it not only includes asymptotic height (Niklas, 1995), but it also shows the transition between neighbour and bending effect (Kohyama *et al.*, 1990; Weiner & Thomas, 1992). The height of the transition point was estimated by the slope of a tangent (*t*) to the curve equaling 1 (Sumida *et al.*, 1997). Nonlinear curve functions suggested by van Laar & Akça (2007) were tested. The best fit function was selected based on the fitness of observed and estimated data, which was indicated by their correlation value (R).

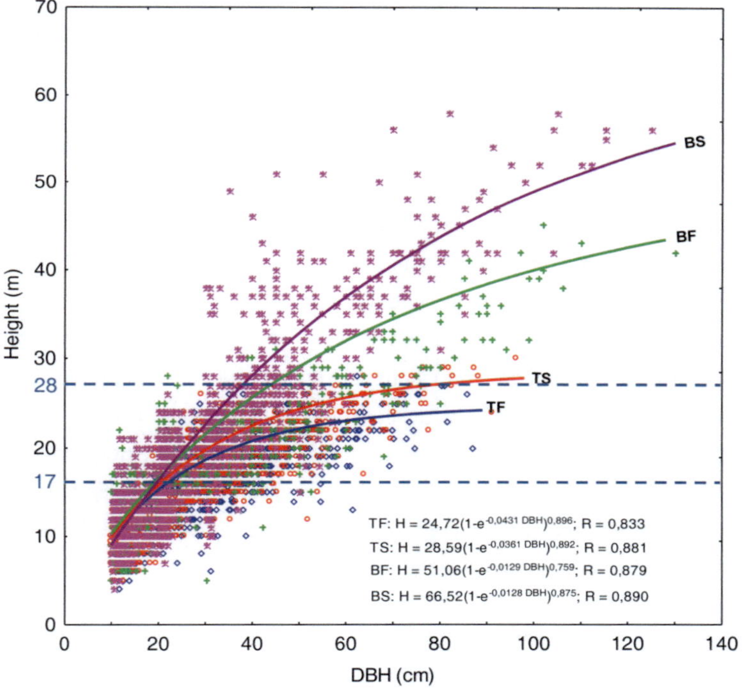

Figure 4.3. Diameter-height relationship of all stems with DBH≥10cm in Tunas flat (TF), Tunas steep (TS), Bonggo flat (BF), and Bonggo steep (BS) stands

Chapman-Richards' growth function fitted the observed data well (R > 0.8) in all stands (Figure 4.3). Not all curves reached a plateau, but the curves of TF and TS stands were closer to their asymptotes and fell earlier below those of BF and BS stands. These trends indicated that biomechanical constraints affected canopy trees of the Tunas stand more than those in the Bonggo stands; the mechanical force's effect was more robust in the upper storey of the flat stand in both forests. Competition for light resources among trees in the lower canopy was less intense in the Tunas stands than in the Bonggo stands, as can be seen from their transition points of 17m and 28m in height. This figure was supported by the density of trees in the middle storey. In the Tunas forest, trees preferred to gain an advantage from shading each other in this storey rather than exposing their crown to light. It can also be seen that their numbers were not significantly different between flat and steep stands; in the Bonggo forest, trees that benefited from light were most likely to enter the middle storey after winning a competition in a series of gap formations. The number of these successful trees was significantly higher in the flat stand. As a result, many trees were under pressure from the neighbour effect in the lower storey.

4.4. Seedling and sapling characteristics of the primary forest

4.4.1. Species composition

The purposes of this study were (1) to describe regeneration patterns (species compositions, diversities, and abundance) within each of two topographic conditions (flat and steep slopes) in the Tunas and Bonggo forests of Papua-Indonesia, (2) to identify species similarities between regeneration and canopy trees within each of the four primary forest stands and (3) to assess the structural characteristics of seedlings and saplings in each of four stands.

Data analyses were conducted as follows: First, the importance value (IV) was used to describe and compare species composition of study stands. IVs of seedlings and saplings were the total of relative density and relative frequency (IVI = RD+RF) of each tree species, as elaborated in Sabogal (1992). RD is the relative density (%) of a species, which reflects the proportion of individuals that belong to this species compared to the stand's total individuals. RF (%) is the relative frequency of a species, which explains the ratio of plot(s) where the particular species is found to the total plots in a study stand. The total IV for all tree species in one stand is 200. The ten most essential species were ranked based on their IV and indicated separately. The remaining species were pooled together and represented by one summed IV. Second, three indices described the four study stand's diversity, including Simpson's index, Shannon-Wiener index, and Shannon's evenness measure. Third, the degree of species similarity between the canopy and the regeneration storeys was assessed using Sorensen's index of similarity described in McCune *et al.* (2002).

The ten most dominant species of regenerated trees based on IV rank of four stands in Tunas and Bonggo primary forest are presented in Table 4.12 and 4.13, respectively. As can be seen from the tables, species numbers, abundances, and frequency of regenerating tree species varied within and among stands, though the closed-distance stands (flat and steep sites) were more similar than the far-distance stands (Tunas and Bonggo forests). As a comparison, the seedlings of steep areas within both forests were about two species higher than in the flat areas, while the saplings were five species more in the steep area than the flat area. In contrast, there were 10 to 11 species at the seedling stage, and 14 species at the sapling stage encountered more in Tunas forest than those recorded at the Bonggo plots.

Absolute abundance and frequency of certain common species at seedling and sapling layers, as shown by their important values (IV), were also relatively different between neighbouring stands, and were substantially different between distant forests. In the Tunas forest seedling IV ranking (Table 4.12), *Vatica rasak* was the first, followed by *Litsea timoriana* in the second place of the flat stand. With the highest IV rank of 21.61% at this level, *Vatica rasak* had the largest individual ha^{-1} (3,750) and frequency distribution (60%). *Litsea timoriana*, in the second place, had an IV of 17.06%, 3,250 individuals ha^{-1}, and was distributed in less than 40% area of the stand. The third place was *Garcinia celebica*, with an IV of 16.30%, and the fourth place was *Blumeodendron amboinicum*, with an IV of 12.64%.

At the sapling level of The Tunas flat stand, *Vatica rasak* was still ranked first, with the highest IV of 14.66%, 520 individuals ha^{-1}, and evenly distributed in a wider area (90%) than its seedlings. As in the seedling layer *Litsea timoriana* followed it in ranking, with an IV of 10.23%, 320 individuals ha^{-1}, and distributed over 70% of the area. Like the seedlings, the third and the following ranks of saplings were mostly represented by the same species (6 out of 8 species except *Chisoheton ceramicus* at the fourth place and *Canarium hirsutum* at the tenth place) but with different sequence rank order.

Table 4.12. The ten most important species of seedlings and saplings ranked based on species' Important Value (IV) at two topographic stands in **Tunas** primary forest.

Seedlings			Saplings		
No.	Species	IV (%)	No.	Species	IV (%)
Flat forest stand					
1	*Vatica rassak*	21.61	1	*Vatica rassak*	14.66
2	*Litsea timoriana*	17.06	2	*Litsea timoriana*	10.23
3	*Garcinia celebica*	16.30	3	*Blumeodendron amboinicum*	8.77
4	*Blumeodendron amboinicum*	12.64	4	*Chisocheton ceramicus*	8.40
5	*Whiteodendron papuana*	12.25	5	*Pometia pinnata*	7.31
6	*Pometia pinnata*	11.75	6	*Whiteodendron papuana*	7.10
7	*Palaquium obtusifolium*	9.98	7	*Syzygium versteegii*	6.73
8	*Syzygium versteegii*	8.59	8	*Garcinia celebica*	6.52
9	*Myristica argentea*	6.82	9	*Palaquium obtusifolium*	6.37
10	*Syzygium anomalum*	6.32	10	*Canarium hirsutum*	6.16
	Others (23) 62%	76.68		Others (36) 41%	117.75
	Total species (33)	200		Total species (46)	200
Steep forest stand					
1	*Vatica rassak*	26.38	1	*Vatica rassak*	19.47
2	*Litsea timoriana*	21.16	2	*Litsea timoriana*	11.76
3	*Garcinia celebica*	12.26	3	*Blumeodendron amboinicum*	9.55
4	*Chisocheton ceramicus*	12.26	4	*Chisocheton ceramicus*	8.56
5	*Blumeodendron amboinicum*	9.89	5	*Garcinia celebica*	8.00
6	*Canarium asperum*	9.89	6	*Syzygium anomalum*	7.57
7	*Syzygium anomalum*	9.89	7	*Myristica argentea*	6.92
8	*Garcinia dulcis*	7.09	8	*Myristica globosa*	6.59
9	*Canarium indicum*	6.60	9	*Calophyllum inophyllum*	6.44
10	*Calophyllum inophyllum*	5.66	10	*Myristica tubiflora*	6.35
	Others (24) 61%	78.92		Others (41) 45%	108.81
	Total species (34)	200		Total species (51)	200

Table 4.13. The ten most important species of seedlings and saplings ranked based on species' Important Value (IV) at two topographic stands in **Bonggo** primary forest.

	Seedlings			Saplings	
No.	Species	IV (%)	No.	Species	IV (%)
	Flat forest stand				
1	*Homalium foetidum*	18.63	1	*Homalium foetidum*	13.63
2	*Celtis rigescens*	17.64	2	*Celtis rigescens*	12.56
3	*Anisoptera polyandra*	14.48	3	*Anisoptera polyandra*	9.96
4	*Canarium asperum*	13.49	4	*Calophyllum papuanum*	8.88
5	*Litsea ledermannii*	13.49	5	*Litsea ledermannii*	8.88
6	*Palaquium obtusifolium*	12.44	6	*Pometia pinnata*	8.88
7	*Calophyllum papuanum*	12.44	7	*Canarium indicum*	8.66
8	*Canarium australianum*	11.39	8	*Syzygium anomalum*	8.66
9	*Syzygium anomalum*	10.40	9	*Canarium asperum*	8.12
10	*Pometia pinnata*	9.34	10	*Canarium australianum*	8.12
	Others (13) 67%	66.27		Others (22) 48%	103.66
	Total species (23)	200		Total species (32)	200
	Steep forest stand				
1	*Celtis rigescens*	21.86	1	*Celtis rigescens*	16.28
2	*Anisoptera polyandra*	19.26	2	*Anisoptera polyandra*	14.17
3	*Homalium foetidum*	14.57	3	*Calophyllum inophyllum*	13.55
4	*Calophyllum inophyllum*	13.27	4	*Homalium foetidum*	12.67
5	*Canarium asperum*	13.27	5	*Canarium asperum*	12.31
6	*Litsea ledermannii*	13.27	6	*Syzygium anomalum*	11.18
7	*Pometia pinnata*	11.58	7	*Diospyros sp*	10.20
8	*Syzygium anomalum*	11.58	8	*Myristica sulcata*	9.58
9	*Syzygium versteegii*	11.58	9	*Calophyllum papuanum*	9.32
10	*Myristica sulcata*	10.28	10	*Litsea ledermannii*	9.32
	Others (15) 70%	59.48		Others (27) 59%	81.42
	Total species (25)	200		Total species (37)	200

In the Tunas steep stand, the composition pattern was more similar to the flat stand, which may be because they shared about 60% of the species (detail is in the Appendix). Both stands also had the same first three most important species in the IV rank: *Vatica rasak, Litsea timoriana,* and *Garcinia celebica* from the top to the third rank at the seedling layer, and *Vatica rasak, Litsea timoriana,* and *Blumeodendron amboinicum* in the same order at the sapling layer.

Although they remained at the same highest positions, the composition of those important species at separate regeneration layers was different between flat and steep stands in terms of absolute abundance and frequency. This was an indication that species responded differently to their environment. Some species increased their quantity and/or frequency, but others did inversely. For instance, the IVs of *Vatica rasak* showed a similar increase at seedling and sapling stages from flat to steep stand (by 4.77% and 4.81%, respectively), whereas the IV of *Litsea timoriana* indicated inconsistent increases of 4.10%, and 1.5% at the seedling and sapling levels, respectively. These IV increases related to the stems of *Vatica rasak* and *Litsea timoriana* were added by 250 and 500 individuals ha^{-1} at the seedling layer. By 320 and 100 individuals ha^{-1} at the sapling layer in the steep stand. These increasing individuals were mainly from existing (dominant) species of seedlings and saplings. These denominators' changes contributed to the higher IV of several species in steep stands as they gained a higher proportion relative to other species. Other important species that had their IVs increase from flat to steep stand were *Chisocheton ceramicus, Canarium asperum, Syzygium anomalum, Garcinia dulcis, Canarium indicum,* and *Calophyllum inophyllum.* In contrast, some important species had their IVs reduce from flat to steep stands, such as *Garcinia celebica, Blumeodendron amboinicum, Whiteodendron papuana, Pometia pinnata, Palaquium obtusifolium, Syzygium versteegii,* and *Myristica argentea.* These species constantly had the same pattern of composition change at the seedling as at the sapling layer.

Although mostly different species composed the Bonggo forest, the common species that responded variously to flat and steep stands was most likely similar to those of the Tunas forest.

In the Bonggo forest, the most dominant species in the flat stand was *Homalium foetidum,* with the highest IVs of 18.63% and 13.63% at seedling and sapling layers. Its IVs were lower in the steep stand, being 14.57%, and 12.67%, respectively. It concerned reducing its abundance from 2,000 stems ha^{-1} (seedling) and 280 stems ha^{-1} (sapling) to 1,500 stems ha^{-1} and 240 stems ha^{-1}, respectively. The frequency distribution also revealed a drop by 10% at the seedling layer and 20% at the sapling layer from flat stand to steep stand. In contrast, *Celtis rigescens*, which was the second-most important species in the flat stand, with 17.64% IVs at the seedling storey and 12.56% at the sapling storey, increased to 21.86% and 16.28% at these respective storeys. It was also supported by an increasing frequency of 20% at the seedling layer and a constant higher frequency distribution of 80% in the sapling layer. Other species that had their abundance and frequency increased from flat to steep stand were *Cheltis rigescens, Anisoptera polyandra, Calophyllum inophyllum, Pometia pinnata, Syzygium anomalum, Syzygium versteegii,* and *Myristica sulcata.* The species that reduced their abundance and frequency from flat to steep stand were *Canarium asperum, Litsea ladermannii, Palaquium obtusifolium, Calophyllum papuanum,* and *Canarium australinum.* These species had the same pattern in both layers, except for *Pometia pinnata* and *Canarium indicum.*

4.4.2. Species diversity

This study presents diversity indices of regenerating species of flat and steep stands of Tunas and Bonggo forests in Tables 4.14 and 4.15. As depicted in the tables, the diversities of lower-canopy tree species of all stands were considered high. The seedling richness of Tunas flat and steep areas was 33 and 34 species, and the sapling richness of these stands was 46 to 51 species, respectively. Species abundance was significantly lower in Bonggo forest, ranging from 23 to 25 species of seedlings and 32 to 37 species of saplings.

Table 4.14. Diversity measures of lower-canopy tree species of two topographic stands in **Tunas** primary forest.

Diversity measures	Flat stand		Steep stand	
	Seedling	Sapling	Seedling	Sapling
Species richness	33	46	34	51
Shannon (H')	3.12	4.54	3.08	3.47
Shannon maximum (H' $_{max}$)	3.50	3.83	3.53	3.93
Shannon evenness (E)	0.89	0.93	0.87	0.88
Simpson (1-D)	0.95	0.97	0.94	0.96

Table 4.15. Diversity measures of lower-canopy tree species of two topographic stands in **Bonggo** primary forest.

Diversity measures	Flat stand		Steep stand	
	Seedling	Sapling	Seedling	Sapling
Species richness	23	32	25	37
Shannon (H')	2.97	3.31	2.95	3.24
Shannon maximum (H' $_{max}$)	3.14	3.47	3.22	3.61
Shannon evenness (E)	0.95	0.95	0.92	0.89
Simpson (1-D)	0.95	0.97	0.95	0.96

All forest stands showed high diversity in lower-canopy tree species. The Shannon E indices were all high (over 0.87), and the Simpson 1-D indices were even higher (> 0.94). Nevertheless, there was a reduction in the diversity of seedlings and saplings from flat to steep stand. H' decreased from 3.12 (seedlings) and 4.54 (saplings) in the flat to 3.08 and 3.47 in the steep stand. Also, the evenness (E) of 0.89 (seedlings) and 0.93 (sapling) in the flat stand dropped to 0.87 and 0.88 at the respective regeneration layers of the steep stand. The decrease in Simpson diversity (1-D) at the Tunas forest's steep stand was hardly noticeable, since the Simpson index is weighted toward the dominant species.

The same trend in the diversity decreases of the Tunas forest was also encountered in the Bonggo forest. Shannon E showed more reduction of diversity in the saplings than in the seedlings as the number of rare species increased three-fold in the sapling layer. In contrast, the decline of the Simpson index between the respective stands was hardly noticeable.

The slight reduction in diversity indices is related to the rising number of rare species in the steep stand. Shannon indices (diversity H' and evenness E) are highly considered proportion of rare species. As the species with 1 or 2 individual(s) increased in the steep stand, the proportion of species skewed with longer tail resulted in reducing Shannon diversity and evenness. On the other hand, the Simpson index emphasizes the arithmetic mean of all species' proportional abundance, which is less sensitive to rare species.

4.4.3. Regeneration of canopy tree species

The percentages of canopy tree species with regeneration at lower-canopy layers in Tunas and Bonggo stands were estimated using Chao-Jaccard abundance-based estimation for similarity index (Table 4.16 and 4.17).

As illustrated in Table 4.16, the percentage of canopy tree species with available regeneration in the Tunas' steep stand was higher than those in the flat stand. Chao-Jaccard Indices indicated that 60% and 67% or 52 out of 93 species of the canopy trees in the steep stand had their regeneration in seedling and sapling layers, respectively. Compared to the flat stand, the indices of similarity were significantly lower (52% and 57% at seedling and sapling layers, respectively) as the upper-layer trees shared about 50 of 104 species. These lower indices could reflect smaller species richness of seedling and sapling (43 and 51 species) in the flat stand than those (50 and 59 species) found in the steep area.

Table 4.16. Chao-Jaccard abundance-based indices of similarity between seedling, sapling, and canopy tree species in flat and steep stands of **Tunas** primary forest.

Comparing tree layers	Estimated no. of species in			Chao-Jaccard similarity index (Percent)
	First layer	Second layer	Shared layers	
Flat stand				
Seedling-Sapling	43	51	38	73
Seedling-Canopy	43	104	51	52
Sapling-Canopy	51	104	50	57
Steep stand				
Seedling-Sapling	50	59	50	88
Seedling-Canopy	50	93	52	60
Sapling-Canopy	59	93	52	67

In the Bonggo forest, the percentage of canopy tree species with regeneration also varied slightly between flat and steep stands (Table 4.17). At the seedling layer, the canopy species' availability in the flat area, as revealed by the Chao-Jaccard similarity index, was 5% lower than that shown in the steep site. 20 out of 68 canopy species were encountered at the flat stand's seedling layer, whereas 30 of 73 canopy species were counted at the same tree category in the steep site. Furthermore, the species similarity index of sapling-canopy was recorded inversely, 4% higher in the flat stand as canopy shared its 29 from 68 species to sapling in the flat stand and 50 from 73 species sapling in the steep stand. It was unclear whether the variation of similarity indices in this forest could be explained by species richness, but overall there were higher numbers of canopy species that shared their number with seedlings and saplings in the steep than flat stand.

Table 4.17. Chao-Jaccard abundance-based indices of similarity between seedling, sapling, and canopy tree species in flat and steep stands of **Bonggo** primary forest.

Comparing tree layers	Estimated no. of species in			Chao-Jaccard similarity index (Percent)
	First layer	Second layer	Shared layers	
Flat stand				
Seedling-Sapling	24	33	23	80
Seedling-Canopy	24	68	20	60
Sapling-Canopy	33	68	29	72
Steep stand				
Seedling-Sapling	35	47	33	80
Seedling-Canopy	35	73	30	65
Sapling-Canopy	47	73	50	68

Chao-Jaccard Index estimated the species similarity not only based on species richness alone but also on their abundances. Species succession to steep stand allowed the increase of species number; then, the completion became tighter. As a result, some species reduced their abundance when they reached the sapling stage, thus increasing this layer's richness.

4.4.4. Regeneration abundance

The natural regeneration densities in Tunas and Bonggo forests were robust at seedling and sapling layers, showing potential for future forest growth, as illustrated in Tables 4.18 and 4.19.

The mean number of regenerate trees in the Tunas forest (Table 4.18) in the seedling category was 28,250 individuals ha^{-1} in the flat stand and 26,750 individuals ha^{-1} in the steep area. In comparison, in the sapling category, there were 5,480 individuals ha^{-1} in the flat area, and 6,080 individuals ha^{-1} in the steep stand. Their standard errors of less than 10% indicated high precision of the sampling method.

Table 4.18. The abundance of natural regeneration in flat and steep stands of **Tunas** primary forest.

Slope of stands	Seedling			Sapling		
	n.ha^{-1}	SD	SE (%)	n.ha^{-1}	SD	SE (%)
Flat	28,250	2,898.76	10.26	5,480	518.12	9.45
Steep	26,750	2,648.38	9.90	6,080	534.99	8.80

Wilcoxon Matched Pairs Test	Wilcoxon Matched Pairs Test
Valid N = 8	Valid N = 10
T = 11.0000	T = 4.0000
Z = 0.9802	Z = 2.3953
P = 0.3270[ns]	P = 0.0166[*]

The paired comparison demonstrated statistically that seedling abundance did not differ significantly between flat and steep stands of Tunas forest; however, sapling abundance showed a significant difference between both forest communities, according to the Wilcoxon Matched Pair (WMP) Test (p=0.0166).

Table 4.19. The abundance of natural regeneration in flat and steep stands of **Bonggo** primary forest

Slope of stands	Seedling			Sapling		
	n.ha^{-1}	SD	SE (%)	n.ha^{-1}	SD	SE (%)
Flat	23,750	2,429.56	10.23	3,720	269.98	7.26
Steep	19,250	2,058.18	10.69	3,240	350.24	10.81

Wilcoxon Matched Pairs Test	Wilcoxon Matched Pairs Test
Valid N = 9	Valid N = 9
T = 0.0000	T = 2.0000
Z = 2.6656	Z = 2.4286
P = 0.0077[**]	P = 0.0151[*]

In the Bonggo forest, the average number of stems in the seedlings category had a slightly broader range, between 19,250 individuals ha^{-1} in the steep stand and 23,750 individuals ha^{-1} in the flat stand (Table 3.19). The range was in agreement with the results of the WMP test, which indicated that both stands showed a highly significant difference in their seedling abundance (p=0.0077). Also, the average number of stems in the sapling category was narrower, ranging from 3,240 individuals ha^{-1} in the steep stand to 3,720 individuals ha^{-1} in the flat stand. However, the WMP test showed a significant difference of sapling abundance between the two stands (p=0.0151). The standard errors above 10%, except the mean sapling abundance of the flat stand, were still acceptable for a high-accuracy sampling method.

4.4.5. Diameter and height of saplings

The mean diameter of saplings calculated in arithmetic and quadratic equations, as well as the mean height of saplings in arithmetic and Lorrey's formulas for all stands in Tunas and Bonggo forests, are presented in Table 4.20.

Table 4.20. Diameter and height of saplings at flat and steep stands in Tunas and **Bonggo** forests

Slopes	Diameter		Height	
	Arithmetic (mean±SD,cm)	Quadratic (mean, cm)	Arithmetic (mean±SD,m)	Lorrey (mean, m)
Tunas Forest				
Flat	3.03 ± 0.16	3.71	3.80 ± 0.19	6.68
Steep	3.13 ± 0.22	3.83	3.85 ± 0.21	7.13
WMP-test	ns	-	ns	-
Bonggo Forest				
Flat	3.59 ± 0.33	4.10	4.16 ± 0.31	6.40
Steep	3.99 ± 0.39	4.22	4.56 ± 0.49	7.20
WMP-test	ns	-	ns	-

WMP-test = Wilcoxon Matched Pairs Test; ns = no significant difference

Statistically, the mean diameters of saplings at flat and steep stands in both forests showed no significant difference based on the Wilcoxon Matched Pair test (WMP-test). It seemed that steep stands showed a higher mean diameter of 3.13 cm in the Tunas forest and 3.99 cm in the Bonggo forest, compared to 3.03 cm and 3.59 cm of the flat stands in the respective forests. This implies that individual saplings inhabiting the steep stands were slightly larger in diameter. This inference was also supported by the presence of the same trend in the quadratic diameter of respective stands, which weighted the mean diameter towards larger-diameter sapling populations. As displayed in Table 4.20, quadratic-based calculation sifted the arithmetic mean diameters by approximately 0.70 cm bigger at both stands (flat and steep) in Tunas Forest. The sifted quadratic diameter was large enough compared to the sifted diameter of 0.51 cm at the flat stand, and even larger than the 0.23 cm diameter at the steep stand in the Bonggo forest. This indicated that the Bonggo forest was composed of more saplings with big diameters than the Tunas forest, especially at Bonggo's steep forest.

In further regard to the vertical structure, Table 4.20 revealed that there was no significant difference between the height of saplings of the flat and steep stand within both forests. However, there was a marked difference in the mean heights of saplings between both forests. The sapling heights of Tunas were lower than those of Bonggo. For comparison, sapling heights of Tunas ranged between 3.80m and 3.85m, whereas those of Bonggo spanned from 4.16m to 4.56m.

In addition to Lorrey's mean height, the steep stands showed slightly higher values than those shown by the flat stands in both forests. However, Tunas and Bonggo forests had similar Lorrey's mean heights. As illustrated by Table 4.20, the Lorrey's mean heights of the Tunas forest saplings were 6.68 m and 7.13 m at the flat and steep forests, respectively. Thus, the respective heights in the Bonggo forest were 6.40 m and 7.20 m. Lorrey's mean weights heights toward basal areas. Accordingly, it is useful to evaluate the growth of a stand. In this case, the results indicated better under-canopy growth in both forests' steep stands.

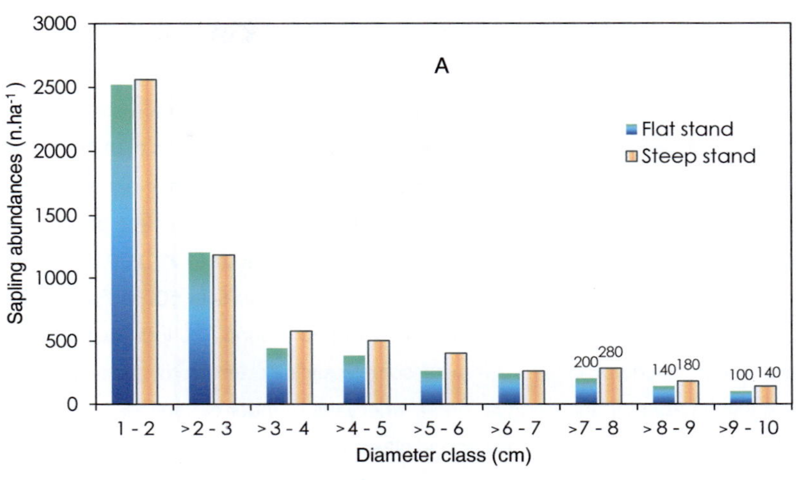

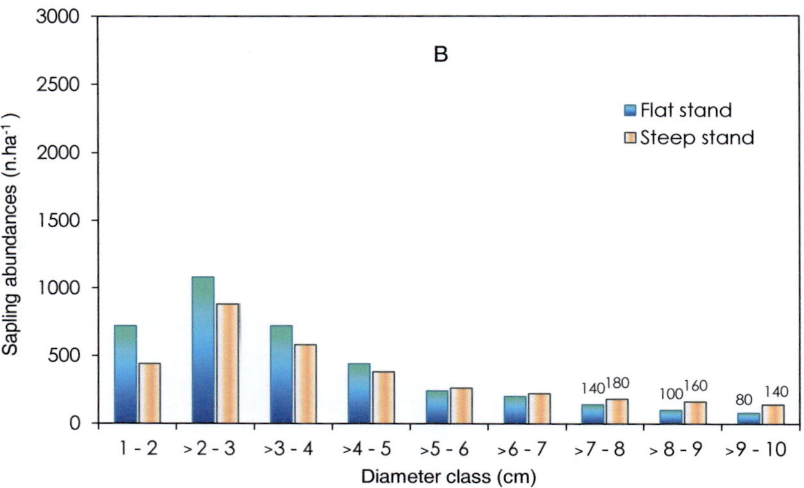

Figure 4.4. Sapling distribution by diameter class at flat and steep stands in Tunas forest (A) and Bonggo forest (B)

Another finding was the proportion of sapling abundance in diameter size classes (Figure 4.4). The graphs showed that all stands had a general trend in sapling distribution by diameter size of a decreasing proportion of saplings with increasing diameter classes. Even though Tunas and Bonggo forests indicated highly abundant saplings in the smaller classes (less than 3 cm in diameter), they exhibited differences in reducing proportional abundance to the next higher diameter classes. For example, in the Tunas forest, the number of saplings dropped suddenly from about 2500 stems ha^{-1} in the first class to 1200 individuals ha^{-1} in the second class, and then reduced moderately to over 500 individuals ha^{-1} in the third class, before gradually decreasing to around 120 individuals ha^{-1} in the 10-cm diameter class. This was completely different from the abundance reduction pattern in the Bonggo forest, where the number of saplings had a peak of about 1000 stems ha^{1} in the second class, then gradually decreased to around 100 saplings ha^{-1} in the highest diameter group. This sudden decrease implied an interruption in the regeneration growth of the Tunas forest. It is perhaps high recruitment of juveniles to tree level (DBH>10cm). On the other hand, gradual reduction indicates a continuous supply of regeneration stocks in the Bonggo primary forest because of the slow growth of understorey individuals. In the topographic effect, the numbers of saplings in big diameter classes (mainly above 5 cm) at steep stands were higher in proportion than that in the respective classes at the flat stand. This means that the steep stands in both forests showed a more successful succession of regenerate stems to the higher-diameter classes. Many juveniles successfully grew, indicated by their large size, because of more light availability in the more sloping areas.

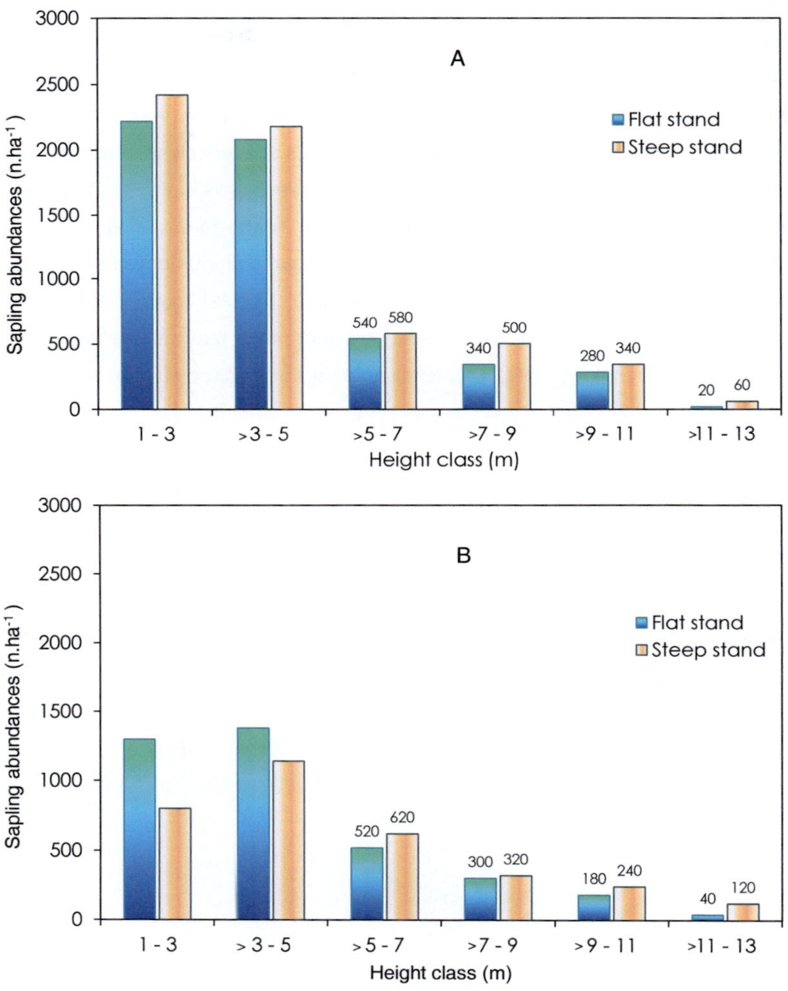

Figure 4.5. Sapling distribution by height class at flat and steep stands in Tunas forest (A) and Bonggo forests (B)

Further findings in sapling distribution and vertical direction (Figure 4.5) show that all stands had a high number of saplings populating the height classes of less than 5 meters. In the Tunas forest, the juveniles' population in these classes was remarkably higher than in the Bonggo forest; the number in Tunas was more than double that in Bonggo. However, at the height groups greater than 5 meters, all the stands indicated the same number of individuals at every class. This revealed that the chance of juvenile establishment in both forests was high. Still, the higher number of smaller height class individuals in the Tunas forest implied that the saplings had more opportunities to access the resource (light) necessary to begin their development. The probability of successful tropical forest regeneration primarily depends on some level of natural disturbance that causes disordered canopy gaps (Denslow, 1987), i.e., small openings in a canopy (full or partial tree fall) permitting light transmission to seedlings and juveniles.

Moreover, even though the Tunas forest had many established individuals, only about 20% could attain a higher level (>5 m). This sudden selection might be related to the other resources that limit the saplings' density at the higher development stage. Another reason could be the neighbouring effect, which causes some species to intercept others' growth due to more illumination availability at that height level. In this case, the light-loving species would accelerate their vertical attainment rather than horizontal growth. This is likely, considering that mid- and upper-canopy tree species mainly rely on the gap-light for their regeneration (Denslow, 1987; Kuusipalo *et al.*, 1997; Struhsaker, 1997).

5. Logged forests of Tunas and Bonggo

5.1. Tree species composition in unlogged and logged forests

The Indonesian selective logging system for natural forest (TPTI) allows all harvestable commercial timbers, with DBH $\geq$ 40, 50 or 60 cm depending on the type of forest, to be extracted in a cutting cycle of 35 years. In TPTI, the forestry ministry determines the annual allowable cut (AAC) for a logging concession area after assigning the targeted area and evaluating the potential volume of harvestable timber size trees, excluding 20% that are designated as future seed trees. Accordingly, the harvesting intensity is variable and depends on the density of existing harvestable commercial trees. In Borneo, where primary lowland forests exhibit a high density of cuttable trees (23 stems ha^{-1} DBH > 50 cm and 16 stems ha^{-1} DBH > 60 cm), TPTI can result in highly variable felling intensities from 1 to 17 stems ha^{-1} involving various cutting techniques that cause 26% to 46% tree damage (Sist et al., 1998).

A logged forest is considered to be in a degraded condition if, at a certain threshold of timber harvest intensity level (when reaching a specific number of extracted trees ha^{-1}), its damage completely prevents non-pioneer trees from regenerating naturally (Struhsaker, 1997; Sist & Nguyen-Thé, 2002; van Gardingen et al., 2006). Sist et al. (1998) indicate that a maximum of eight trees ha^{-1} must be a limit for felling intensity, regardless of non-reduced or reduced impact logging in Borneo.

In Papua, the harvest intensities varied among the timber concession areas. For example, in the Tunas Timber Lestari (TTL) concession in Tunas forest, the designated AAC is 36.77 m^{3}ha^{-1}, equivalent to 22 harvestable commercial stems ha^{-1} with DBH $\geq$ 40 cm (RKU PT. TTL, 2009), whereas in the Wapoga Mutiara Timber (WMT) concession in Bonggo forest, the AAC is 54.97 m^{3}ha^{-1} or about 16 trees ha^{-1} (RKU PT. WMT, 2012). This AAC of the two logging concessions is very high compared to the limit of 8 trees ha^{-1} that guarantees a sustainably logged forest (Sist et al., 1998). It was expected, therefore, that the logging practices in both concession areas have destroyed and killed the remaining trees, thus completely preventing the growth of non-pioneer commercial timbers and eventually degrading the forests' potential production and diversity. However, little information on these impacts was available. Therefore, this study aimed to analyse tree species composition, diversity and structure in the logged stands and to predict the forests' continuing timber production and species diversity conservation based on an evaluation of the short-time response of species following logging.

Data used in this logged forest study were collected from eight 1-ha plots, or two plots each, from 4-year logged and 8-year logged stands in the logging concessions of PT. TTL in Tunas and PT. WMT in Bonggo (Sections 2.1 and 2.2 give detailed information about the two concession areas). In addition, data from four 1-ha plots of primary stands (further called unlogged stands) in respective logging concessions were used as a comparison in data analysis.

The plot arrangement and design, tree measurement, tree category and data analysis are explained in chapter 2.

More tree species at canopy level (i.e. stems with DBH > 10 cm) were counted in Tunas forest than in Bonggo forest for all (unlogged and logged) stands, with 117 and 84 species, respectively. They were then classified into three commercial types as defined by the companies: high-value commercial (H), mixed commercial (M) and non-commercial (N) timber species.

The Tunas stands were composed of 23 (19.7%) high-value timber species (H), 41 (35.0%) mixed commercial timber species (M) and 53 (45.3%) non-commercial timber species (N). The Bonggo stands consisted of 29 (34.5%) H species, 29 (34.5%) M species and 26 (31.0%) N species. This finding revealed that both locations are highly diverse in commercial timber species; in particular, Bonggo forest contained significantly higher diversity in its high-value timber species than Tunas forest.

Figure 5.1a. Tunas unlogged- (A) and logged-forest 4 years after logging (B)

Figure 5.1b. Tunas logged-forest 8 years after logging (C)

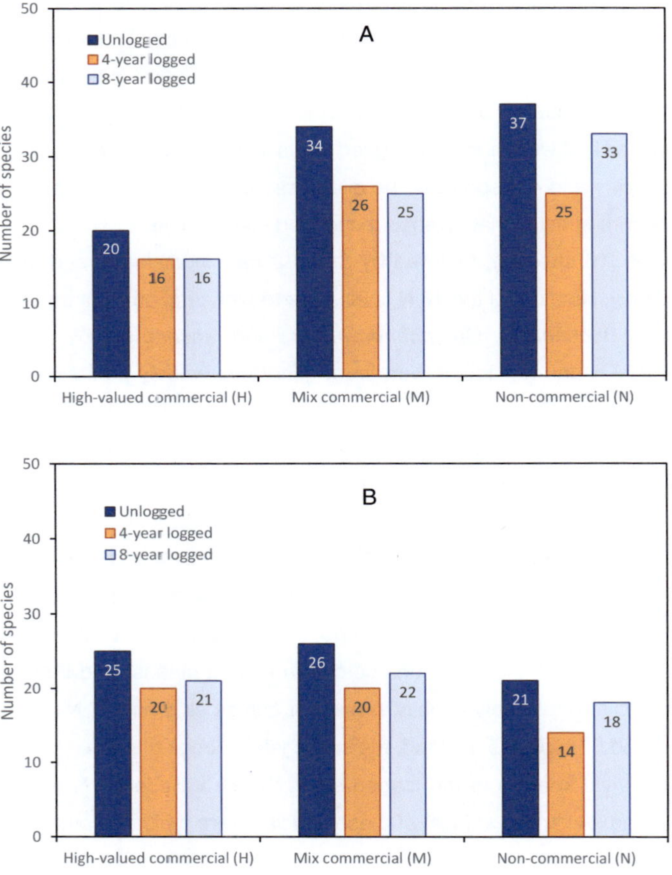

Figure 5.2. Numbers of high-value, mixed and non-commercial tree species (DBH > 10 cm) recorded in unlogged, 4-year and 8-year logged stands in Tunas (A) and Bonggo (B) forests.

The graphs of the distribution of commercial-type species in Tunas and Bonggo forests (Figure 5.2) show that the number of species in all groups is reduced markedly in the logged stands, but only the N species could significantly recover in 8 years. The N species also showed the highest reduction and recovery in species numbers. Comparing the two forests, larger changes (reduction and recovery) in species counts occurred in the Tunas forest. The study found that the lost species and the new arrival species in the logged stands were mostly those with individuals < 3 stems or rare species. In the Tunas unlogged stand, 34 rare species that were absent in logged areas included 19 N species, followed by 11 M species and 4 H species. The disappeared species were then replaced by 21 other rare species that invaded the logged stands. These consisted mostly of N species (16 species), followed by 3 and 2 new incoming H and M species, respectively. Nine out of the 16 N species were pioneers, such as those from the genera of *Buchanania*, *Gomphandra*, *Ficus* and *Macaranga*. By contrast, the commercial H and M species were from genera commonly present in unlogged stands such as *Canarium*, *Palaquium*, *Cananga* and *Cerbera*. The pioneers' invasion, therefore, caused a significant increase in the number of N group species in the 8-year logged stand of Tunas forest.

In the Bonggo forest, the alteration of rare species between the intact stand and the stands disturbed by logging was not as large as that shown in the Tunas forest. The degradation in species number involved a loss of 16 rare species from the unlogged stand that were unrecorded in logged stands, and these species were distributed relatively equally in number across all groups:7 N species, 6 M species and 5 H species. Instead, the space left by logging allowed the growth of 10 new arrival species in the logged stand, which included 5 N species, 3 M species and 2 H species. Most of these species were typically common in intact forest.

Further analysis of the stand composition was performed to rank the species based on their Importance Value Indices (IVIs). The species ranks for unlogged and logged forests can be found in the Appendix. Here, the study lists only the ten top-ranked species for each stand (Tables 5.1 and 5.2).

Table 5.1. Ten most important canopy tree species (DBH > 10 cm) ranked based on the Importance Value Index (IVI), recorded in unlogged, 4-year and 8-year logged stands of **Tunas** forest

Groups	Species	Absolute Density (n/ha)	Frequency (%)	Dominance (m²/ha)	IVI (%)
	Unlogged forest				
H	*Vatica rassak*	38	93	3.47	35.81
M	*Litsea timoriana*	17	85	1.05	16.34
N	*Blumeodendron amboinicum*	10	55	0.76	10.41
H	*Anisoptera polyandra*	7	43	1.04	10.03
N	*Teijsmanniodendron bogoriense*	8	48	0.85	9.90
N	*Pimelodendron amboinicum*	8	48	0.59	8.70
H	*Canarium asperum*	8	45	0.56	8.41
H	*Hopea papuana*	7	45	0.68	8.38
N	*Beilschmiedia* sp.	7	48	0.58	8.08
M	*Syzygium versteegii*	7	40	0.47	7.28
	Other species (77)	152		11.39	176.67
	Total species (87)	269		21.45	300
	4-year logged forest				
H	*Vatica rassak*	29	90	1.72	22.16
M	*Litsea timoriana*	23	65	0.70	13.19
N	*Blumeodendron amboinicum*	22	65	0.62	12.47
H	*Canarium asperum*	17	65	0.51	10.64
N	*Pimelodendron amboinicum*	17	60	0.55	10.59
M	*Syzygium versteegii*	17	55	0.51	10.05
N	*Teijsmanniodendron bogoriense*	14	70	0.37	9.51
M	*Myristica sulcata*	14	45	0.60	9.41
N	*Beilschmiedia* sp.	14	55	0.47	9.26
H	*Canarium indicum*	12	45	0.58	8.85
	Other species (54)	291		10.01	183.87
	Total species (64)	468		16.64	300

Groups	Species	Absolute			IVI (%)
		Density (n/ha)	Frequency (%)	Dominance (m²/ha)	
	8-year logged forest				
H	*Vatica rassak*	53	90	2.43	26.34
M	*Litsea timoriana*	28	90	0.94	14.26
H	*Canarium asperum*	22	80	0.79	11.89
N	*Blumeodendron amboinicum*	20	90	0.66	11.38
M	*Myristica sulcata*	22	65	0.57	10.13
H	*Canarium indicum*	16	70	0.67	9.68
N	*Pimelodendron amboinicum*	14	75	0.63	9.38
M	*Syzygium versteegii*	14	70	0.46	8.19
H	*Palaquium obtusifolium*	13	50	0.59	7.76
N	*Teijsmanniodendron bogoriense*	12	65	0.46	7.67
	Other species (62)	313		12.17	183.32
	Total species (72)	526		20.37	300

The results show that six to seven of the ten most dominant species were commercial timber trees (H and M species) in unlogged and logged stands of Tunas (Table 5.1). The high-value timber species (H) *Vatica rassak, Anisoptera polyandra, Canarium asperum* and *Hopea papuana* made up the highest proportion (total IVI of 62%) of the ten dominant species in the unlogged forest of Tunas. At the same (unlogged) stand, the mixed timber species (M) *Litsea timoriana* and *Syzygium versteegii* accounted for 23.62%. The non-commercials *Blumeodendron amboinicum, Teijsmanniodendron bogoriense, Pimelodendron amboinicum* and *Beilschmiedia* sp. made up the remaining 37.09%.

The investigation also revealed that the IVI of the H species (41.65%) in the 4-year logged stand was markedly lower than that of the unlogged stand. *V. rassak* contributed significantly to this drop (if the logged stand's original condition resembled that of the undisturbed stand). The IVI of *V. rassak* declined from 35.81% to 12.16% following the four-year cutting. This reduction was related mainly to the marked difference of *Vatica*'s basal area (BA) between the two stands; the BA of *Vatica* in the 4-year logged stand was 50% less than that of the unlogged stand. The other high-value species, such as *A. polyandra* and *H. papuana,* showed a trend similar to that of *V. rassak* but not *C. asperum.* The first two species had higher numbers of big trees and fewer small stems. When big trees were removed from the stand, *A. polyandra* and *H. papuana* showed a significant reduction in big individuals and thus showed very low BA; they were then ranked outside of the top ten dominant species in the 4-year logged stand. *C. asperum,* however, showed a very high number of small stems and smaller numbers of large trees. Removing its big trees and creating canopy gaps might spur growth of young individuals, resulting in increasing IVI of *C. asperum.*

In contrast, the IVI values of M species (32.75%) and N species (41.83%) were higher in the 4-year logged stand than in the unlogged stand. *Litsea timoriana, Syzygium versteegii and Myristica sulcata* (M species) represented different preferences in timber demand. *L. timoriana* was the second most important species after *V. rassak.* Its individual distribution was identical to that of *V. rassak,* i.e. many individuals in both small and big stem classes. Its 50% dominance (BA) in the 4-year logged stand, lower than that in the unlogged stand, indicated that the harvestable stems of *L. timoriana* were most likely in high demand. In comparison, *S. versteegii* and *M. sulcata* showed increasing dominance (BA) and an increasing number of (smaller) individuals four years after logging because, as these trees were less in demand timber species, the big trees of *S. versteegii* and *M. sulcata* still remained and the small trees benefited from the canopy opening in the logged stand. Accordingly, the decreasing dominance of *L. timoriana* (high-demand timber) was offset by the increasing growth of lower-demand *S. versteegii* and *M. sulcata,* resulting in the higher IVI of mixed commercial species in the 4-year logged stand. In addition, like less-profitable commercial species, the N species such as *B. amboinicum, P. amboinicum, and T. bogoriense* also benefited from the available space to increase their BA, even though these non-commercial species decreased in the number of individuals, perhaps due to mortality of small-sized trees caused by collateral effects of harvesting.

The findings also indicated that the high-value timber species had recovered significantly in the eight years after logging. The summed IVIs of these species increased from 41.65% in the 4-year logged stand to 55.67% in the 8-year logged stand. When comparing the absolute numbers of individuals and dominances between 4- and 8-year logged stands, the individual recruitments (into the smallest diameter class) were more significant than the increment of BA (in big trees). Recovery was also recorded in the mixed and non-commercial dominants, but the individuals and basal area increments of the last two timber species groups were not as large as those of the high-value timber group. This slower recovery was clearly reflected by the small addition of individuals and BA of the mixed and non-commercial dominants in the 8-year logged stand. Even though the stems and BA of dominants from the two groups increased, their proportions among all species in the stand (in terms of IVIs) were reduced to 32.58% and 28.43%, respectively.

In the Bonggo stands, nine of the ten most important species were the commercial species, both in unlogged and logged stands (Table 5.2). The high-value timber species (H) *Intsia bijuga, Homalium foetidum, Celtis rigescens, A. polyandra,* and *Pometia pinnata* contributed the highest proportion of IVI, 90.72%, in the unlogged forest of Bonggo. The next largest proportion belonged to the mixed commercial (M) group (IVI of 34.89%), such as *Litsea ledermannii, M. sulcata, Myristica tubiflora* and *Syzygium anomalum. P. amboinicum* was the only non-commercial (N) species among the ten dominants, with an IVI of 9.06%.

The results indicated that H species showed a significant reduction in IVI from 90.72% in the unlogged stand to 63.49% in the 4-year logged stand. *I. bijuga* contributed the greatest IVI reduction (20%), while others were responsible for the remaining 7% reduction. *I. bijuga* had ten big-sized stems with a total BA of 4.77 m^{2}ha^{-1} in the unlogged stand. However, the species remained with very scarce stems (3 ha^{-1}) and small BA (0.12 m^{2}ha^{-1}) in the 4-year logged stand, indicating that *I. bijuga* was in very high demand.

Table 5.2. Ten most important canopy tree species (DBH > 10 cm) ranked based on the Importance Value Index (IVI), recorded in unlogged, 4-year and 8-year logged stands of **Bonggo** forest

Group	Species	Absolute Density (n/ha)	Frequency (%)	Dominance (m²/ha)	IVI (%)
		Unlogged forest			
H	*Intsia bijuga*	13	0.69	4.77	23.24
H	*Homalium foetidum*	33	0.74	1.72	18.20
H	*Celtis rigescens*	36	0.70	1.46	17.83
H	*Anisoptera polyandra*	25	0.68	2.00	16.93
H	*Pometia pinnata*	19	0.60	1.83	14.52
M	*Litsea ledermannii*	19	0.56	1.21	12.23
M	*Myristica sulcata*	21	0.48	0.81	10.72
M	*Myristica tubiflora*	20	0.46	0.88	10.68
M	*Syzygium anomalum*	19	0.50	0.79	10.26
N	*Pimelodendron amboinicum*	16	0.41	0.79	9.06
	Other species (66)	225		14.79	156.33
	Total species (76)	448		31.06	300
		4-year logged forest			
H	*Homalium foetidum*	20	0.80	1.32	19.18
H	*Celtis rigescens*	19	0.70	0.97	16.31
M	*Litsea ledermannii*	17	0.70	0.93	15.27
H	*Anisoptera polyandra*	15	0.65	1.04	15.14
H	*Pometia pinnata*	11	0.45	1.10	12.86
N	*Pimelodendron amboinicum*	14	0.50	0.65	11.53
M	*Syzygium anomalum*	16	0.55	0.34	10.66
M	*Myristica tubiflora*	12	0.60	0.46	10.27
M	*Mastixiodendron pachyclados*	10	0.60	0.50	9.92
M	*Myristica sulcata*	11	0.55	0.45	9.76
	Other species (44)	177		9.31	169.10
	Total species (54)	320		17.07	300

Group	Species	Absolute			IVI (%)
		Density (n/ha)	Frequency (%)	Dominance (m²/ha)	
	8-year logged forest				
H	*Homalium foetidum*	37	0.95	2.73	27.74
H	*Celtis rigescens*	39	0.75	1.13	19.96
M	*Litsea ledermannii*	22	0.70	1.23	15.65
M	*Syzygium anomalum*	22	0.70	0.93	14.31
H	*Anisoptera polyandra*	16	0.70	1.07	13.25
M	*Myristica sulcata*	16	0.90	0.58	12.50
H	*Canarium asperum*	20	0.65	0.67	12.22
H	*Pometia pinnata*	12	0.55	1.12	11.59
M	*Myristica tubiflora*	12	0.65	0.63	9.93
N	*Pimelodendron amboinicum*	13	0.35	0.55	7.93
	Other species (51)	176		12.13	154.93
	Total (61)	383		22.77	300

Other dominants, including H, M and N species, also showed a slight drop in stem numbers and BA in the 4-year logged stand but not as outstanding as that of *I. bijuga*. The ten dominants were more affected by logging since they lost about 8.5 m^2ha^{-1} of BA compared to the 5.48 m^2ha^{-1} lost by the less dominant and rare species together, represented by 66 species.

All species in the 8-year logged stand showed an increase in their total BA by 5.70 m^2ha^{-1} compared to the 4-year logged stand (Table 5.2). The top ten dominants accounted for more than half of the increment in BA (2.88 m^2ha^{-1} or 51%), and the other 51 species (less dominant and rare species) contributed less than half of the BA recovery (2.82 m^2ha^{-1} or 49%). Out of the ten dominants, only three H species contributed significantly to BA recovery (2.29 m^2ha^{-1} or 80%): *H. foetidum* (1.41 m^2ha^{-1}), *C. rigescens* (0.16 m^2ha^{-1}) and *C. asperum* (0.30 m^2ha^{-1}). On the other hand, the BA of other H species (*A. polyandra* and *P. pinnata*) remained constant between four and eight years following logging.

The dominant H species also showed a remarkable increase in absolute numbers of individuals from the 4-year to 8-year logged stands; they accounted for the majority (94%) of 64 added individuals. This significant increase in stems confirmed that the BA increment of the dominant H species did not reflect the diameter growth of existing adults (big trees) but mostly represented young recruitments. However, among the H species, the number of young trees increased in *H. foetidum, C. rigescens* and *C. asperum* 8 years after logging, but not in *I. bijuga, A. polyandra or P. pinnata.* The increase in abundance also occurred in M species such as *L. ledermannii, S. anomalum* and *M. sulcata,* but not in *M. tubiflora.* Species' light requirements may explain their different abundance recovery eight years after harvesting.

The above analysis reveals that the Tunas and Bonggo forests varied greatly in intact stand composition and logged stand recovery after disturbance as indicated by their significantly different numbers and types of species with various specific characteristics.

The Tunas stand, which typically contained more rare species, experienced more species loss than the Bonggo stand did during logging activity, because species with few stem numbers might be more sensitive to deletion from a community. Cannon *et al.* (1998) argued that logging has less impact on species reduction because, instead of vanishing from the stand, rare species may be present in the stand, but they are not counted in sample plots due to their sparse distribution. Many rare species not only could occur locally (e.g. at stand level) but also often occur extensively in a biogeographical range and may represent a vast absolute number of individuals at the landscape level (Primack & Hall, 1992; Poorter *et al.,* 1996; Pitman *et al.,* 1999). This may be true because, in this study, there were 34 rare species in the unlogged stand that were not recorded in the logged stand of the Tunas forest, and 21 rare species that occurred in the logged but not unlogged stands, showing rapid species recovery in a short period after logging. Significantly fewer rare species were present in Bonggo's unlogged stands and absent in its logged stands (16 species) or absent in Bonggo's unlogged stands and present in its logged stands (10 species) compared to Tunas. Most of these rare species are non-commercial timbers, and some of them are well-known as late-successional species, which are commonly present in the lower canopy. However, the appearance of early successional species in the Tunas logged stand implies that a new succession has begun that could gradually change the future species composition.

Most of the dominant species of both forests were hardwood commercial timbers. According to Richards (1996), most of the high-value timber species have long life spans and slow growth rates. Based on species composition analysis, these tree species responded differently to logging disturbance. Some responded quickly in the eight years after logging by producing many young recruitments, such as *V. rassak, C. asperum, L. timoriana and S. versteegii* in Tunas, and *H. foetidum, C. rigescens, C. asperum, L. ledermannii, S. anomalum* and *M. sulcata* in Bonggo. Even though they were cut, these species were abundantly available for future advanced growth because they were most important in their respective forests. Mawazin & Subiakto (2013) stated that the most important species are more stable because they are more adaptable to the environment than rare species. Other commercial timber species were slowly recruited or seldom found among young trees, but adults were still available for the next cutting rotation because they were protected through a requirement to protect the future crop trees. These species included *A. polyandra, H. papuana* (in Tunas), *I. bijuga* (in Bonggo) and many less important commercial species in both forests. The absence of young commercial specimens would allow the existing adults with lower timber volumes to be harvested in the second cutting cycle and potentially become extinct timber species in third-cycle harvesting. Therefore, in many cases, most of the timber concession areas are limited to no more than one cycle period of 30–35 years, compelling the timber companies to re-enter the same places for the second and third cutting cycles (Zimmerman & Kormos, 2012).

5.2. Tree species richness and diversity in the logged forests

5.2.1. Species richness

Species richness (S) was markedly higher in Tunas forest than in Bonggo forest, and it was also higher in unlogged stands than in logged stands of both forests, but the trend was inconsistent when similar comparisons were made to the individual–species (I–S) ratio (Table 5.3). The unlogged stand of Tunas had significantly higher species richness (91 species) than did the intact stand in Bonggo (72 species). The richness of these undisturbed stands dropped remarkably in logged stands, with species numbers tending to increase again from the more recently logged stand (4 years) to the earlier-logged (8 years) stand.

Table 5.3. Species richness and individual–species ratio of canopy trees at three stands (unlogged, 4-year logged and 8-year logged) in Tunas and **Bonggo** forests

Measure of richness	Forest stands		
	Unlogged	4-year logged	8-year logged
		Tunas	
Species richness (S)	91	67	74
Individual–species (I–S) ratio	3.32	7.16	7.27
		Bonggo	
Species richness (S)	72	54	61
Individual–species (I–S) ratio	11.93	6.17	6.49

The increasing I–S ratio in the Tunas 4-year logged forest was consistent with the increase in its total trees; the stems increased by 199 ha^{-1} from the unlogged to the 4-year logged stands (see Table 5.1). The ten dominant species contributed 30% of the total additional stems or about 12 stems per species. The less important (64) species were responsible for nearly three additional stems per species. The other 23 species were not listed and were responsible for the richness reduction after logging; they were mostly represented by three or fewer individuals, with the possibility to be recorded or unrecorded during sampling.

The decrease of the I–S ratio in the Bonggo 4-year logged forest occurred in relation to its total density reduction. Table 5.2 shows 128 fewer stems in the 4-year logged stand than in the unlogged stand. The ten dominant species contributed 80 stems (63%) of the total stem reduction. Among them, the five top-ranking species were the high-value timbers, and four of these (*I. bijuga*, *H. foetidum*, *C. rigescens* and *A. polyandra*) were responsible for the loss of more than 13 stems per species (or above the average of eight stems). The less important (66) species, on average, accounted for about one stem reduction each, while the other 20 species (mostly rare) did not appear in the 4-year logged forest.

Even though the tree species richness had started to increase after eight years in both forests, the I–S ratio remained stable in both forests at this point. This was because the species added to the richness counting most likely came from rare species or species with fewer individuals. The rare species could represent both the existing slow-growing species that could be under- and over-sampled and the fast-growing pioneer species that offset the richness reduction.

5.2.2. Species diversity indices

It is unclear which groups of species (common or rare) contributed more to the forest diversity if the measure is based on individuals per species, because the ratio refers to the total number of stems and species in a specific area. Therefore, heterogeneity was used to explain the distribution of species abundance in more detail; heterogeneity is a two-dimensional measure of diversity in which commonness and rareness of species in a community are emphasised. Diversity indices of canopy trees at three stands in Tunas and Bonggo forests (unlogged, 4-year logged and 8-year logged) are presented in Table 5.4.

Table 5.4.   Diversity indices of canopy trees at three stands (unlogged, 4-year logged and 8-year logged) in Tunas and Bonggo forests

Measure of diversity	Forest stands		
	Unlogged	4-year logged	8-year logged
Tunas			
Simpson's diversity (1-D)	0.97	0.98	0.97
Shannon–Wiener Diversity (H')	3.97	3.86	3.88
Shannon Exp. Diversity ($e^{H'}$)	52.98	47.47	48.42
Shannon Evenness (E)	0.89	0.92	0.90
Bonggo			
Simpson's diversity (1-D)	0.97	0.97	0.96
Shannon–Wiener Diversity (H')	3.75	3.64	3.59
Shannon Exp. Diversity ($e^{H'}$)	42.52	38.09	36.23
Shannon Evenness (E)	0.87	0.91	0.87

The slightly higher evenness indicated by the Simpson 1-D and Shannon E values in both logged forests provided evidence that there was a relative increase in the proportion of equally common species after logging, regardless of increasing or decreasing numbers of individuals. In Tunas forest, 64 less-important species contributed about 70% of added stems, while the ten dominants were responsible for 30%. In the unlogged forest, less important species had a range of individuals from 3 to 6 per species, while dominant species ranged from 7 to 38 stems per species (Table 5.1). Gaps created after logging caused many less-important species to increase their numbers of individuals (more than 7 per species) to approach those of the dominants. This alteration in the species composition of the Tunas logged forest demonstrates that more species in the forest became equally common. However, in Bonggo's logged forest, 63% of the stem reduction (or on average eight individuals per species) did affect the dominant species, whereas the 44 less-important species accounted for the rest (37%), with an average loss of 1 stem per species. In the unlogged forest, the individuals of less-important species reached less than 16 stems per species, while the numbers of dominants ranged between 16 and 38.

The top five dominants (all high-value commercial species) had > 25 individuals per species. A sharp reduction in individuals of the dominants resulted in more species with equal proportions in abundance.

Tunas and Bonggo forests also showed a small diversity reduction in 4-year logged stands as indicated by the Shannon H and e^H indices; these Shannon measures related the diversity to the rareness. A decline in the number of rare species in both forests most likely contributed to the slight drop in tree diversity. After eight years, tree diversity had increased in Tunas but not in Bonggo forest.

The following are ten rare commercial species under-sampled in the logged forest of Tunas: *Anisoptera thurifera*, *Cinnamomum culilawan*, *H. foetidum*, *Dacriodes* sp., *Diospyros* spp., *Manilkara fasciculata*, *Podocarpus amara* and *Podocarpus nerlifolius* (high-value commercials), and *Cananga odorata* and *Calophyllum costatum* (mixed commercials). The 13 rare commercial species under-sampled in Bonggo's logged forest are *A. thurifera*, *Canarium rigidum*, *Canarium* sp., *Canarium molucense* and *Cinnamomum culilawan* (high-value commercials), and *Buchanania macrocarpa*, *Dillenia alata*, *Duabanga moluccana*, *Dysoxylum amooroides*, *Eucalyptus deglupta*, *Garcinia dulcis*, *Sterculia parkinsonii* and *Swietenia macrophylla* (mixed commercials).

Overall, the logging impact did not significantly reduce the degree of diversity in Tunas and Bonggo forests. However, the recovery of diversity had been achieved shortly after logging (8 years) only in the Tunas forest but not in the Bonggo forest.

These results are consistent with the statement that in a community with high species diversity, the existence of rare species (which have few or single individuals in plot samples) is susceptible to rarefaction, which is a reduction in density by random deletion of individuals (Heck Jr *et al.*, 1975; Denslow, 1995). If reducing individuals from the unlogged to the logged stand were equivalent to rarefaction alone, the diversity would not change because richness is positively related to stem density (as shown in Tunas forest). Alternatively, Cannon *et al.* (1998) stated that damages and ecological changes associated with logging influenced some species negatively (or positively), thus changing their relative abundances; an alteration in I–S ratio and species per unit area would be expected (as shown in Bonggo forest).

5.3. Stand structure of logged forest

5.3.1. Diameter distribution

The tree abundance of each DBH class was plotted against the DBH class to illustrate the pattern of diameter distributions for each stand. The width of each DBH class was 10 cm. Tree abundances in every diameter class (y-axes) were plotted against the centre of a class interval of 10 cm (x-axes), as shown in the bar graphs in Figures 5.3 and 5.4. An exponential equation $N=a \cdot e^{-b \cdot x}$ was then applied to fit the observed tree distribution. N is the number of trees in a diameter class, e is Euler's number (2.71828), a is a constant number limiting the maximum number of trees at the smallest diameter since DBH > 0, and b is a constant describing the slope of the curve. A matrix correlation was used to test the fit of observed and estimated values at $p=0.05$ and a confidence interval (CI) of 95%. The results indicated that estimated values of all stands fitted the observed values well with a significant relationship ($p < 0.05$) at CI 95%, and coefficients of correlation (R) of all stands were above 0.986, indicating that 98% of the estimated numbers of stems for every class were explained well by the model.

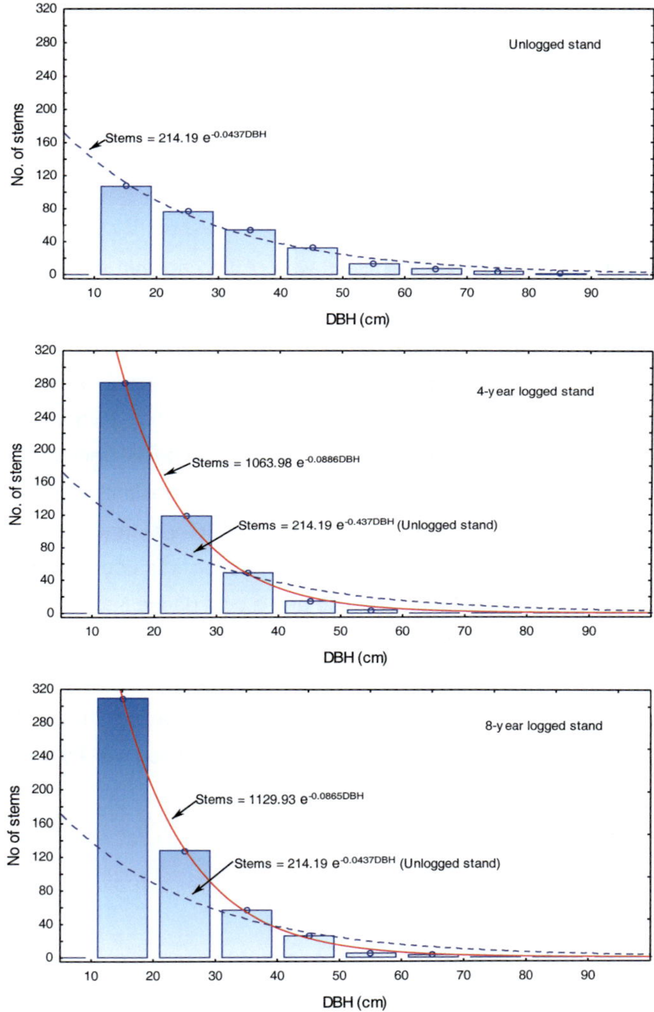

No. of stems: Observed (bar); predicted for unlogged forest (dash line);
Predicted for logged forest (solid line)

Figure 5.3. Observed and predicted numbers of stems in 10-cm DBH size class intervals in unlogged, 4-year and 8-year logged stands in **Tunas** forest.

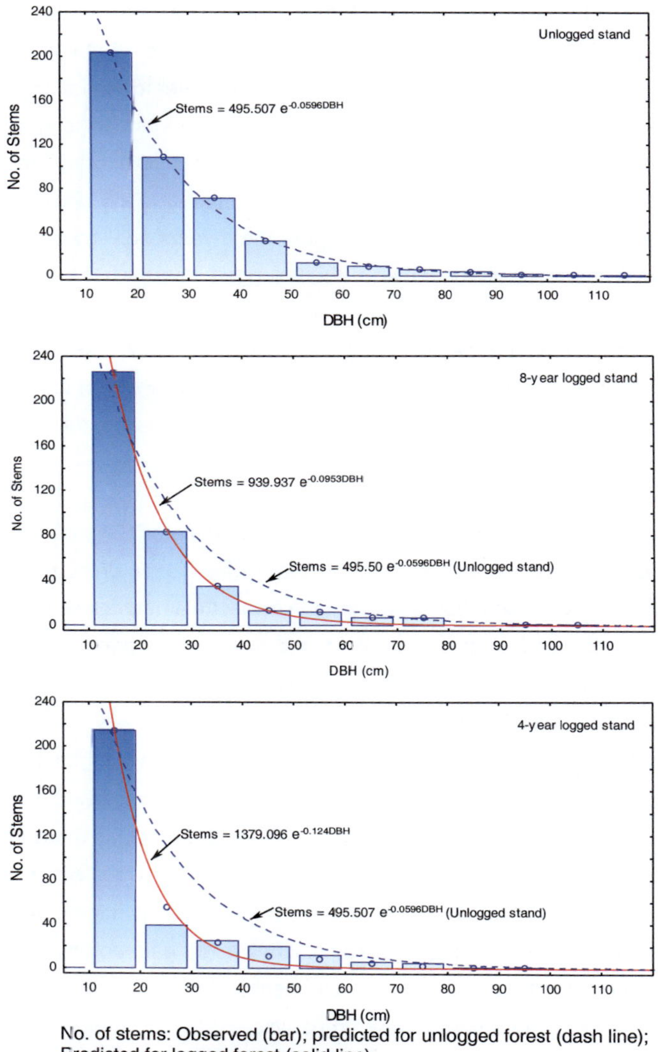

Figure 5.4. Observed and predicted numbers of stems in 10-cm DBH size class intervals in unlogged, 4-year and 8-year logged stands in **Bonggo** forest.

It can clearly be seen from Figure 5.3 that a gradual decrease in tree numbers from about 100 stems in the smallest class (10–20 cm) to 78 stems in the next class to one or two stems in the maximum diameter class of 90 cm was characteristic for the unlogged stand of the Tunas forest. In logged stands, however, the curves were significantly different. About 50% of all stems were concentrated in the smallest diameter class. Compared to the unlogged stand, the number of stems in the smallest class of the 4-years ago logged stand was three times the respective number in the unlogged stand, while the number of stems in the same class of the 8-years ago logged stand was more than three times the number in the unlogged stand. Both logged stands also had about 40 stems more in the second diameter class than did the unlogged stand. Maximum diameter trees were found in the 60–70 and 70–80 cm classes, respectively, for the 4-year and 8-year logged stands.

Meanwhile, the number of stems in the smallest diameter class of the Bonggo unlogged stand (Figure 5.4) was double that of the Tunas unlogged stand. A dramatic drop from about 200 stems in the smallest diameter class to 96 stems in the next diameter class is shown for the Bonggo unlogged stand. This is followed by a gradual decrease to a maximum diameter size class of 110–120 cm. After tree harvesting, the 4-year logged stand shows heavy destruction as indicated by a large difference in the number of stems in the category that should be left for future cutting (20–50 cm diameter size class) compared to that of the unlogged stand. The 20–30 cm size class suffered the most, as more than 50% of trees from this class were killed by collateral damage. Trees of the smallest size in the 4-year logged stand, however, were slightly higher in number compared to those in the unlogged stand. They might gain an advantage from recruited regeneration due to the canopy opening. In the older logged stand (8 years), the number of smallest-sized trees was still higher than in the younger logged stand, indicating regeneration recruitment was still in progress. Many of the stems from this class were also recruited to the next class (the most damaged class in the 4-year logged stand), resulting in positive recovery, but a longer duration might be expected.

5.3.2. Species distribution in diameter size classes

Apart from non-commercial species, the numbers of commercial species (H and M species) and their stems had a peak in the diameter classes of 20-40 cm in unlogged stands (Table 5.5 and 5.6). It was explained in section 4.3.6 that trees of these diameter ranges were in the transition between preferentially height and horizontal (diameter) growth with a height of 17m in Tunas and 28m in Bonggo due to constraining neighbour and biomechanical effects. Most species can grow up and accumulate to this point as they get a chance to light from small canopy gaps or shading and biomechanical protection from neighbours. However, only certain species with light-demanding traits and strong biomechanical character can pass the point to reach and emerge from the canopy layer.

Therefore, the species numbers dropped dramatically from the transition point to the next diameter size class, and decreased gradually to the bigger classes. In both forests, these selected species were mostly represented by commercial species. There were 21 and 30 commercial species, respectively, in Tunas and Bonggo, that could successfully grow above the transition point, and only eight and 13 species that reached the minimum harvestable size (DBH>50cm). Each species was usually represented by less than two stems ha^{-1}, except *Intsia bijuga* that was represented by 3 to 5 stems ha^{-1} in the Bonggo forest.

Despite having more stems and species above harvestable size, the Bonggo forest also had a higher volume of commercial stocks than the Tunas forest. The Bonggo stand had numerous stems at small diameter size classes allowing many trees to grow up to the higher transition point of 28m (than that of 17m in Tunas), before the biomechanical constraint balanced the height with wider diameter size. Tables 5.5 and 5.6 imply that the Bonggo forest had a higher volume of commercial timber than Tunas, since it had a higher transition growth and more trees with big diameter size up to 125cm. In comparison, the Tunas forest had fewer trees with the diameter ranging up to 90cm.

Table 5.5. The number of stems and species distributed in five size classes of the three timber commercial groups of **Tunas** stands

Species	Stands	<20	20-40	40-50	50-60	60-80	>80	Total	<20	20-40	40-50	50-60	60-80	>80	Total
					Stems ha^{-1}							*Species ha^{-1}*			
H	UL	30	48	14	6	6	1	104	17	19	12	4	5	1	20
	L4	72	44	6	2	-	-	123	13	15	6	2	-	-	16
	L8	55	80	10	2	2	-	148	12	14	9	2	2	-	16
M	UL	35	45	10	4	4	1	100	27	29	9	2	3	1	34
	L4	109	71	5	2	-	-	187	22	17	5	2	-	-	26
	L8	128	49	9	1	-	-	186	22	22	9	1	-	-	25
N	UL	51	50	9	4	2	1	116	31	27	6	3	2	1	37
	L4	101	54	4	1	-	-	159	22	15	4	1	-	-	25
	L8	126	56	8	2	1	-	193	29	30	8	2	1	-	33

Table 5.6. The number of stems and species distributed in five size classes of the three timber commercial groups of **Bonggo** stand.

Species	Stands	<20	20-40	40-50	50-60	60-80	>80	Total	<20	20-40	40-50	50-60	60-80	>80	Total
					Stems ha^{-1}							*Species ha^{-1}*			
H	UL	81	98	22	9	11	6	226	18	24	18	9	10	5	25
	L4	85	28	6	5	4	1	128	15	10	6	5	4	1	20
	L8	94	50	6	7	8	1	166	14	11	6	7	8	1	21
M	UL	96	83	13	2	3	1	198	20	25	12	2	3	1	26
	L4	91	42	4	2	1	-	139	18	17	4	1	1	-	20
	L8	99	49	5	2	3	1	159	17	18	5	2	3	1	22
N	UL	39	23	5	2	2	-	70	14	14	5	2	2	-	21
	L4	38	10	2	2	2	1	55	9	6	2	2	1	1	14
	L8	33	19	2	3	2	1	59	9	13	2	3	2	1	18

Species groups: H = High-valued commercial, M = Mixed commercial, N = Non-commercial; Stands: UL = Unlogged, L4 = 4-year logged, L8 = 8-year logged.

As depicted in Tables 5.5 and 5.6, the logging practice removed about the same stems ha^{-1} in both stands, namely 18 trees in Tunas and 19 trees in Bonggo. But, it left different timber stocks after the first cutting, with only 9% of the stems in Tunas and 41% in Bonggo as initial harvestable timber stocks. The removed timbers included more than half of the initial commercial species lost in both forests. The harvesting activity also killed the future crops leaving 85% stems from 66% species in Tunas, and 38% stems from 55% species in Bonggo.

Based on TPTI protocol, 20% of the initial stocks should be kept with no concern to the number of species. The timber harvesting in the Tunas stand failed to meet the requirement. It might be because trees in the Tunas forest were too low in volume to financially balance the operation cost such that the company needed to fulfill a certain quantity of timber volume. Furthermore, the company can rely on future tree crops that remain abundant after the first cutting. However, by considering the species growth rate in this stand, the initial timber volumes will not be reached in the second cutting cycle. In the case of the Bonggo forest, the logging activity could keep the requested minimum timber stock for the next harvesting cycle. However, when felling big trees with wide crown it might have killed many future tree crops, as shown in Bonggo's logged stand. With the lower tree density after first logging, it will affect to the lower transition point, thus reducing the timber volume stock in the next cutting.

Based on the above analysis, to prevent the continuation of timber productivity, the study proposes to decrease the harvesting intensity below eight trees per ha. An indirect approach can be applied in both forests, such species-based cutting limit but with enlarging the targeted harvesting area, especially in Tunas forest.

5.4. Regeneration of canopy tree species

The importance values (IVs) of commercial and non-commercial tree species at the seedling and sapling levels were computed and ranked for unlogged and logged forests. Only the ten top-ranked species are listed for each stand in Tunas and Bonggo forests, respectively, in Tables 5.7 and 5.8.

Table 5.7 presents the composition of the ten most important species at the seedling and sapling stages recorded in unlogged, 4-year logged and 8-year logged stands of the Tunas forest. There were no differences in the composition of seedlings before and after logging. The regeneration of commercial species such as *V. rassak, L. timoriana* and *C. asperum* was consistent, as these were among the most dominant species at the seedling level before and after logging. However, logging caused a reduction in the IVs of almost all dominant regeneration species, which showed various recovery rates after logging. The IV of *V. rassak* decreased from 26.38% in the unlogged stand to 18.54% in the 4-year logged stand but then increased to 21.00% in the 8-year logged stand. The IVs of other commercial species from *Callophyllum, Myristica* and *Syzygium* as well as non-commercial species, which were also listed among the ten most dominant species in the unlogged stand, also decreased after logging. But some of the species were more likely to indicate good recovery eight years after logging. Seral species such as *Ficus* and *Gomphandra* appeared in 4-year logged stands. However, their abundance tended to decrease by eight years after harvesting.

Table 5.7. Ten most important regenerated species ranked based on Importance Value (IV), recorded in unlogged, 4-year and 8-year logged forest of **Tunas**.

Seedlings			Saplings		
Group	Species	IV (%)	Group	Species	IV (%)
		Unlogged stand			
H	*Vatica rassak*	26.38	H	*Vatica rassak*	19.47
M	*Litsea timoriana*	21.16	M	*Litsea timoriana*	11.76
M	*Garcinia celebica*	12.26	N	*Blumeodendron amboinicum*	9.55
N	*Chisocheton ceramicus*	12.26	N	*Chisocheton ceramicus*	8.56
N	*Blumeodendron amboinicum*	9.89	M	*Garcinia celebica*	8.00
H	*Canarium asperum*	9.89	M	*Syzygium anomalum*	7.57
M	*Syzygium anomalum*	9.89	M	*Myristica argentea*	6.92
M	*Garcinia dulcis*	7.09	M	*Myristica globosa*	6.59
H	*Canarium indicum*	6.60	M	*Calophyllum inophyllum*	6.44
M	*Calophyllum inophyllum*	5.66	M	*Myristica tubiflora*	6.35
	Others (24)	78.92		Others (41)	108.81
	Total species (34)	200		Total species (51)	200
		4-year logged stand			
H	*Vatica rassak*	18.54	H	*Vatica rassak*	13.16
M	*Litsea timoriana*	15.85	H	*Canarium asperum*	11.60
H	*Canarium asperum*	14.18	M	*Syzygium versteegii*	11.14
M	*Syzygium versteegii*	13.81	N	*Chisocheton ceramicus*	10.68
N	*Chisocheton ceramicus*	12.14	M	*Litsea timoriana*	10.08
N	*Pimelodendron amboinicum*	10.75	N	*Blumeodendron amboinicum*	8.37
M	*Mastixiodendron plectocarpum*	10.10	N	*Gomphandra* sp.	8.05
N	*Blumeodendron amboinicum*	9.08	N	*Pimelodendron amboinicum*	7.59
M	*Myristica sulcata*	9.08	M	*Myristica sulcata*	7.27
N	*Gomphandra* sp.	8.44	N	*Teijsmanniodendron bogoriense*	6.67
	Others (17)	78.03		Others (34)	105.38
	Total species (27)	200		Total species (44)	200

	Seedlings			Saplings	
Group	Species	IV (%)	Group	Species	IV (%)
	8-year logged stand				
H	*Vatica rassak*	21.00	H	*Vatica rassak*	13.80
M	*Litsea timoriana*	15.65	H	*Canarium asperum*	9.01
H	*Canarium asperum*	15.47	N	*Blumeodendron amboinicum*	8.34
M	*Myristica sulcata*	15.47	M	*Litsea timoriana*	8.01
N	*Beilschmiedia* sp.	11.66	M	*Syzygium versteegii*	7.79
N	*Blumeodendron amboinicum*	11.66	M	*Calophyllum inophyllum*	7.57
M	*Myristica globosa*	9.93	N	*Ficus variegata*	7.36
H	*Palaquium obtusifolium*	9.16	M	*Myristica sulcata*	7.34
N	*Teijsmanniodendron bogoriense*	8.98	H	*Canarium indicum*	7.00
M	*Syzygium versteegii*	8.21	H	*Palaquium obtusifolium*	6.22
	Others (19)	72.81		Others (43)	117.56
	Total species (29)	200		Total species (53)	200

The same pattern of composition changes occurred in the sapling stage; however, the most dominant species were still recovering steadily eight years after harvesting. In the unlogged stand, the IVs of *V. rassak* and *L. timoriana* were 19.47% and 11.76%, respectively. These dropped to approximately 13.16% and 10.08% after logging and remained stable even eight years after harvesting. Other species also experienced a similar alteration in composition. However, the increases in abundance and dominance were more significant in commercial species than in non-commercial species after eight years of recovery.

Table 5.8 shows the composition of the ten most important species at seedling and sapling stages recorded in the unlogged, 4-year logged and 8-year logged stands of Bonggo forest. Unlike the Tunas forest, a remarkable change in regeneration composition at the seedling and sapling levels was found in the Bonggo forest in the years after logging.

Table 5.8. Ten most important regenerated species ranked based on the Importance Value (IV), recorded in unlogged, 4-year and 8-year logged forest of **Bonggo**.

Seedlings			Saplings		
Group	Species	IV (%)	Group	Species	IV (%)
Unlogged stand					
H	*Celtis rigescens*	21.86	H	*Celtis rigescens*	16.28
H	*Anisoptera polyandra*	19.26	H	*Anisoptera polyandra*	14.17
H	*Homalium foetidum*	14.57	M	*Calophyllum inophyllum*	13.55
M	*Calophyllum inophyllum*	13.27	H	*Homalium foetidum*	12.67
H	*Canarium asperum*	13.27	H	*Canarium asperum*	12.31
M	*Litsea ledermannii*	13.27	M	*Syzygium anomalum*	11.18
H	*Pometia pinnata*	11.58	H	*Diospyros sp.*	10.20
M	*Syzygium anomalum*	11.58	M	*Myristica sulcata*	9.58
M	*Syzygium versteegii*	11.58	M	*Calophyllum papuanum*	9.32
M	*Myristica sulcata*	10.28	M	*Litsea ledermannii*	9.32
	Others (15)	59.48		Others (27)	81.42
	Total species (25)	200		Total species (37)	200
4-year logged stand					
M	*Syzygium versteegii*	18.37	H	*Anisoptera polyandra*	16.24
M	*Litsea ledermannii*	15.28	M	*Syzygium anomalum*	15.41
H	*Canarium asperum*	14.70	M	*Litsea ledermannii*	11.45
H	*Pometia pinnata*	14.12	M	*Syzygium versteegii*	9.47
H	*Anisoptera polyandra*	12.64	M	*Calophyllum inophyllum*	7.89
M	*Calophyllum inophyllum*	12.64	M	*Myristica sulcata*	7.89
H	*Palaquium obtusifolium*	11.02	N	*Alphitonia macrocarpa*	7.52
N	*Ficus variegate*	9.99	H	*Canarium asperum*	7.49
N	*Alphitonia macrocarpa*	7.93	H	*Canarium indicum*	7.12
N	*Anthocephalus cadamba*	6.32	H	*Palaquium obtusifolium*	7.09
	Others (21)	76.99		Others (32)	102.43
	Total species (31)	200		Total species (42)	200

	Seedlings			Saplings	
Group	Species	IV (%)	Group	Species	IV (%)
		8-year logged stand			
M	*Syzygium versteegii*	16.96	H	*Anisoptera polyandra*	14.97
H	*Palaquium obtusifolium*	15.44	H	*Canarium asperum*	13.07
H	*Canarium asperum*	14.32	H	*Palaquium obtusifolium*	12.09
H	*Pometia pinnata*	11.68	M	*Syzygium anomalum*	11.13
M	*Calophyllum inophyllum*	11.68	H	*Homalium foetidum*	9.69
M	*Litsea ledermannii*	11.29	M	*Calophyllum inophyllum*	9.17
M	*Myristica sulcata*	11.29	H	*Pometia pinnata*	8.68
H	*Anisoptera polyandra*	10.16	M	*Litsea ledermannii*	8.22
M	*Syzygium anomalum*	10.16	M	*Myristica sulcata*	8.22
N	*Alphitonia macrocarpa*	9.04	N	*Alphitonia macrocarpa*	7.27
	Others (19)	77.99		Others (30)	97.48
	Total species (29)	200		Total species (40)	200

C. rigescens and *H. foetidum* were two of the most dominant high-value species regenerating (as seedlings and saplings) in the unlogged stand, but both were not recorded as important regeneration species in the 4-year and 8-year logged stands. In contrast, other commercial species such as *S. versteegii, A. polyandra, C. asperum, L. ledermannii, P. pinnata, Calophyllum inophyllum* and *Palaquium obtusifolium* remained among the most dominant species at the seedling and sapling stages after logging.

Among the most dominant commercial species, *Anisoptera, Syzygium, Litsea, Calophyllum* and *Myristica* appeared as the most abundant species at the sapling level after four years of logging. Others, such as *Canarium, Palaquium, Homalium* and *Pometia*, also composed the most dominant species at the sapling level, but these species recovered later, after eight years of logging. Additionally, some seral species (*Ficus variegate, Alphitonia macrocarpa* and *Anthocephalus cadamba*) appeared in the 4-year logged stand among the most dominant species; however, their abundance decreased in the 8-year logged stand.

6. Light requirements of tree species

6.1. Concept

One of the major factors that allow plants to begin their growth, establish in a biotope and survive among their competitors in a community is their ability to absorb light energy and convert it into reproductive chemical energy (Attridge, 1990). Tree canopies shade each other, causing reduced and modified irradiance for the understories in the forest. The affected tree species respond by adjusting their light intensity requirements, as reflected in their regeneration, development, and survival (Lieffers *et al.*, 1999; Bloor & Grubb, 2003; Mason *et al.*, 2004), causing respective implications for the succession and diversity of their communities (Beaudet *et al.*, 1999). Light is the most crucial driver for tree growth and development in moist tropical forests, where water is not limited and soil nutrients are available. Therefore, specifying the light requirements of tree species is essential to design silviculture systems appropriately.

Tree species' light requirements have been classified using quantitative methods based on differences in diameter increment (Vanclay, 1991; Richards, 1996), physiological criteria (Whitmore, 1989; Poorter & Werger, 1999) and height growth patterns (Condit *et al.*, 1996; Poorter, 1999). Notably, these methods have successfully classified rainforest tree species into three functional groups (Lamprecht, 1989; Poorter, 1999). The first group is the shade-tolerant species group, which tolerates deep shade and proliferates beneath dense forest canopy (Kerstiens, 2001; Govindarajan *et al.*, 2004). The second is the light-demanding species group, which requires a high-light environment for their establishment (Lamprecht, 1989; Hung, 2008) and throughout their entire lifecycle (Lamprecht, 1989). The third group includes semi-shade-tolerant tree species that share certain characteristics of both, shade-tolerant and light-demanding species (also known as the 'opportunistic' group). The aforementioned research practices have used species growth and survival abilities related to light intensity and absorption to classify them into their respective groups. However, the appropriateness of this classification system remains debatable. This is largely due to the temporal fluctuation of light over seasons and years in tropical regions.

Richards (1996) summarised tropical mixed forest regeneration studies worldwide and identified a diameter class distribution pattern for defining light-demanding, shade-tolerant and semi-shade-tolerant species. Rollet (1974) introduced seven types of diameter class distributions using non-linear curves for trees with >10 cm diameter at breast height (DBH) in the Venezuelan forest. A greater number of curve types (seven versus three) infers that more complexity is found when assigning the form of the non-linear curve to various light requirement traits. This is because, in reflecting their regeneration strategy, some species are in a continuum distribution between strikingly polarised shade-tolerant and most shade-intolerant species. Thus, the shape of the non-linear curve cannot differentiate them. Therefore, the shape of a linearising non-linear curve would serve as a better way of implementing a discrete number (Burn, 1964 after Mitlöhner, 1998).

Another important element that influences and limits plant growth, survival and distribution in many terrestrial ecosystems is nitrogen (Cole & Rapp, 1981; Evans, 2001; Dawson *et al.*, 2002). During the growth process, trees and other plants take up nitrate (NH_3^-) and ammonium (NH_4^+) directly from the air (via their stomata) and the soil solution, which are then assimilated into a nitrogen source for the synthesis of amino acids required for plant growth (Ellenberg & Nettels, 2001; Takagi *et al.*, 2010). Pate (1973) detected that one or more nitrogen-rich compounds (e.g., amides, ureides and amino acids) carry the bulk of the nitrogen and exit the roots of some herbaceous plant species through their xylem. However, other plant species indicated no nitrate assimilation in the root system. Additionally, experiments with *Hordeum vulgare* showed genotypic differences in the relative distribution of nitrate reductase (Warner & Huffaker, 1989). It was concluded that different species could exhibit varied nitrate reductase activity in their root and shoot systems, which can cause intra variation in plant nitrogen content. While nitrate reduction in shoots is markedly stimulated under light conditions, such reduction is not strictly light-dependent (Aslam *et al.*, 1979;

Aslam & Huffaker, 1982). Nitrate reductase is a cytoplasmic enzyme, which may be loosely associated with the outer chloroplast membrane (Wallsgrove *et al.*, 1979).

Previous studies have indicated that physiological factors such as different nitrogen uptake mechanisms, different cycles of assimilation and nitrogen recycling in plants can provide new information on the metabolism of nitrogen in various species of higher plants (Pate, 1973). Since the aforementioned factors can discriminate against ^{15}N (Evans, 2001), they can be insightful in nitrogen-stable isotope studies. Although most existing studies were performed in closed systems (lab-oriented), others have detected a wide range of light requirements for plant species using stable nitrogen isotopes in natural systems. This review specifically addresses how the assimilation pathway of nitrogen can influence xylem δ^{15}N. It is expected that variation in sapwood δ^{15}N abundance can explain light response among plants.

The following part of the study presents the role of δ^{15}N and total N in woody tissues in discriminating tree species into light requirement groups. Hence, the goals of this study were as follows: (1) to assess the relationship between the pattern of species' diameter distribution (slope b) and their δ^{15}N or total N content in sapwood tissue; (2) to determine the correlation between species δ^{15}N and total N content; (3) to classify tree species into three light requirement groups based on slope b pattern, δ^{15}N and total N.

This investigation recorded the species names, DBH and total heights of all trees with a DBH ≥10cm and collected three replicated sapwood samples per species in 16 1-ha plots distributed in the primary forests of Tunas and Bonggo. Then, species' b-values as the light requirement variable were calculated, and laboratory analysis for sapwood δ^{15}N and total N in sapwood tissue as species trait variables was conducted. The study then assessed correlations between species' light requirements (slope b) as an independent variable and species' nitrogen contents (δ^{15}N and total N) as dependent variables in a non-linear regression analysis. Species were then grouped into their light demand classes.

6.2. Slope of tree species distribution

Species' diameter size-frequency distributions have been used to identify their responses to light intensity in tropical forests; however, the proxy variable represented by this distribution is less relevant. As a result, difficulties appeared when discriminating transition characteristics between shade-tolerant and light-demanding species (Richards, 1996). In the present study, the linear slope of the size class distribution of 50 species (49% of recorded species) from Tunas and 53 species (69% of species encountered) from Bonggo forest simplify the pattern of distribution curves into a single value (b), as presented in Tables 6.1a and 6.1b.

The number of stems listed at every diameter class bin of 10 cm is the average number of species in each respective class investigated in four 1-ha plots. Based on their stem's distribution (Tables 6.1a and 6.1b), the studied species were differentiated into five groups. Group 1 only included species found in the lower canopy with a ≤20 cm DBH. The next two groups included species that reached the intermediate or upper canopy and had a <50 cm maximum DBH. Group 2 regenerated a high abundance of poles (>100 stems ha^{-1}), while Group 3 only presented a moderate number of poles (<100 stems ha^{-1}). The last two groups were the emergent tree species, which appeared above the upper canopy. Species with abundant regeneration were assigned to Group 4 and those with rare or no regeneration were assigned to Group 5.

The form of the diameter frequency distribution was then described using a hyperbolic equation $n = (\frac{1}{d} \cdot a) + b$, where n is the number of stems in a certain size class diameter, d is the mid-value of the class, and a and b are constants. To reduce missing individuals n in a certain diameter size, a 10-cm range class was used. Additionally, to avoid distortions due to the influence of small diameter classes on excessive inclination angle, the equation was linearised by multiplying the equation with the mid-value diameter (Brun, 1964 after Mitlöhner, 1998), thus, $n \cdot d = a + b \cdot d$. The slope ($b$) of the linear curve was then applied as the species' light-response variables.

Table 6.1a. Frequency of tree diameter classes and slope (b) of their linearised curves for every tree species (DBH >5cm) classified into five groups based on their recruitment and establishment ability in **Tunas**.

Species name	5	15	25	35	45	55	65	75	85	95	Slope (b)
Group 1											
Gnetum gnemon	2	1									0.50
Aquilaria polyantha	1	2									2.25
Endospermum moluccanum	1	2									2.25
Group 2											
Litsea timoriana	440	31	3	2							-67.80
Garcinia celebica	292	4	1	1							-43.10
Canarium asperum	363	27	1	1	1						-39.10
Canarium hirsutum	130	1	1								-31.25
Palaquium lobbianum	118	13	1								-28.25
Syzygium anomalum	192	3	2	1							-27.70
Canarium indicum	178	2	1	1							-25.70
Beilschmiedia sp	135	22	6	3							-18.90
Garcinia dulcis	120	2	2								-17.80
Group 3											
Pometia acuminata	82	7	1	1							-12.05
Pometia pinnata	92	12	8	2	1						-9.40
Whiteodendron papuana	64	6	1	1							-9.20
Planchonella firma	62	8	1	1							-9.20
Litsea irianensis	86	2	-	-	1						-7.29
Litsea tuberculata	52	23	2	-	1						-6.86
Litsea ledermannii	24	1	1	1							-5.15
Blumeodendron amboinicum	58	4	3	2	1						-4.80
Palaquium obtusifolium	22	5	1								-4.25
Aceracium diversivolium	20	1	1								-3.75
Podocarpus amara	20	1	1								-3.75
Vitex pubescens	20	-	1								-3.75
Lithocarpus rufovillosus	29	3	1	1							-3.50
Myristica globosa	24	1	1	1							-2.45
Mastixiodendron plectocarpum	22	2	1	1							-2.30
Spathiostemon javensis	21	2	-	1							-1.96

Table 6.1a continued

Group 3

Myristica lepidota	8	1	1								-0.75
Myristica hollrungii	6	1	1								-0.25
Chisocheton ceramicus	9	3	1	1							-0.20

Group 4

Calophyllum papuanum	46	-	3	1	-	1					-3.52
Vatica rassak	105	10	8	6	4	3	3	3	1		-2.65
Hopea papuana	84	6	2	1	1	-	-	-	1		-2.55
Syzygium hylophilum	72	7	3	2	1	-	-	-	1		-2.24
Calophyllum inophyllum	53	2	2	1	1	-	1				-2.15
Syzygium versteegii	38	2	2	2	1	1					-1.74
Anisoptera polyandra	40	1	2	1	2	-	1				-1.04

Group 5

Xanthostemon crenulatus	-	4	1	-	2	1	-	1			0.16
Dillinea ovalifolia	-	3	-	-	-	1					0.25
Cinnamomum culilawan	-	2	3	3	-	-	1				0.49
Neoscortechinia sp.	2	3	1	1	-	1					0.68
Pimelodendron amboinicum	-	1	2	1	2	1	1				0.91
Anisoptera thurifera	1	1	2	-	1	1	1				0.94
Campnosperma auriculatum	2	1	2	-	1						0.97
Alstonia spectabilis	-	-	1	-	-	-	-	1			1.00
Podocarpus nerlifolius	-	-	-	1	1	1					1.00
Syzygium gustavioides	-	-	-	-	-	-	1	1	-	1	1.00
Gironniera subaequalis	1	4	5	4	1						1.60
Aglaia sp.	2	2	2	2	1	2					1.61

Table 6.1b. The frequency of tree diameter classes and slope (b) of their linearised curves for every tree species (DBH >5cm) classified into five groups based on their recruitment and establishment ability in **Bonggo**.

Species name	Diameter classes (cm)												Slope (b)
	5	15	25	35	45	55	65	75	85	95	105	115	
Group 1													
Ficus variegata	2	1											0.50
Vitex sp.	2	1											0.50
Gnetum gnemon	2	3	1										0.70
Alphitonia macrocarpa	2	4	1										0.70
Linociera oblongifolia	1	2	1										1.00
Buchanania macrocarpa	3	2											1.05
Aquilaria sp.	3	2											1.05
Group 2													
Canarium asperum	398	30	4	3	1								-42.35
Syzygium anomalum	332	8	7	3	1								-32.45
Calophyllum inophyllum	166	3	1	1									-24.05
Canarium indicum	114	1	1	-									-16.90
Calophyllum papuanum	145	4	2	1	1								-13.85
Group 3													
Syzygium hylophilum	63	-	1	-	1								-8.29
Mangifera sp	54	4	1	1									-7.40
Celtis rigescens	47	3	1	1									-5.97
Syzygium sp	64	7	3	2	1								-4.51
Mastixiodendron pachyclados	66	4	3	1	1								-2.95
Dysoxylum amooroides	18	2	-	-									-2.00
Terminalia arborea	31	2	1	2	1								-1.80
Diospyros sp.	20	1	-	1									-1.43
Myristica sp.	47	2	3	-									-1.25
Vatica rassak	11	3	1	1									-0.80
Syzygium versteegii	10	3	2	1									-0.40

Group 4

Species													Value
Homalium foetidum	224	19	6	5	2	1	1						-11.18
Hopea papuana	-	-	4	1	1								-2.75
Anisoptera polyandra	83	10	6	5	2	1	1	1	-	1			-2.68
Pterygota horsfieldii	60	7	3	1	1	-	-	1					-2.08
Eucalyptus deglupta	-	8	3	2	-	1							-1.43
Dracontomelum edule	-	-	11	2	-	-	1	-	2				-1.08
Pometia pinnata	34	5	6	3	3	2	1						-0.93
Chisocheton ceramicus	-	6	-	2	-	1							-0.88
Palaquium obtusifolium	28	4	3	2	1	1							-0.71
Pometia acuminata	8	10	3	2	1	1							-0.70
Litsea ledermannii	15	9	4	3	1	1	-	1					-0.64
Toona sureni	32	1	1	1	1	1	-	1					-0.40

Group 5

Species													Value
Parartocarpus spp	-	2	2	1	-	1	1	1					0.66
Dillenia alata	-	2	2	-	-	1	1						0.56
Podocarpus nerlifolius	-	2	-	1	-	-							0.25
Spondias dulcis	-	-	2	1	1	1							0.25
Pimelodendron amboinicum	2	6	3	3	1	1							0.34
Cinnamomum culilawan	-	3	-	1	-	-	1						0.45
Cananga odorata	-	1	2	1	1								0.75
Endospermum medullosum	-	2	1	1	1	1	1						0.79
Alstonia scholaris	-	1	2	1	-	-	1						0.82
Albizia falcataria	-	1	2	1	-	-	1						0.82
Duabanga moluccana	-	2	1	-	1								0.87
Macaranga beillei	1	2	1	-	1								0.87
Campnosperma auriculatum	1	1	2	-	1	1	1						0.91
Intsia bijuga	-	1	1	2	2	1	1	1	2	1	1	1	0.97
Pterocarpus indicus	-	6	2	2	1	1	1	-	1	1	1	2	0.99
Anisoptera thurifera	1	-	-	-	1	1							1.08
Anthocephalus cadamba	-	1	-	2	1								1.25

As shown in Table 6.1, the diameter class distribution varied greatly among species. However, some species showed similar patterns in their individual size distributions. Here, the patterns of size class distribution are elaborated into five groups. Species belonging to Group 2 had abundant stems of the smallest diameter (rarely observed with a large diameter) and followed an exponential distribution with the curve forming a tall "L-shape". When the curves were linearised using the burn method (Brun, 1969, after Mitlöhner, 1998) the slopes (b-values) were negative and far from zero. Recorded in different forests, the values-b of this group ranged from -67.80 to -17.80 in the Tunas stand and from -42.35 to -13.85 in the Bonggo stand. *Canarium asperum*, *Canarium indicum* and *Syzygium anomalum* were species from this group that were commonly found in both stands. All species from this group were listed as the 10th most important species at the regeneration level and at the canopy level in their respective stands (preview sections). Although they regenerated abundantly in moderate shade, these species reduced in stem number with diameter sizes ≤50 cm. On the other hand, Group 5 species showed a scarcity or lack of individuals in the smallest diameter as well as an equilibrium population structure in some cases. Their size class pattern was an "erratic distribution", with some deficiencies of individuals in small or intermediate diameter classes. The b-values of species from this group were positive and closed to zero, thus implying an equal distribution with slightly more individuals of larger diameter. Among the members of Group 5, *Xanthostemon crenulatus* and *Dillenia alata* had the lowest b-values of 0.16 and 0.56, respectively, in the Tunas and Bonggo forests. Moreover, *Aglaia* sp. and *Anthocephalus cadamba* represented the highest b-values (1.61 and 1.25, respectively) in their respective forests. They were mostly found as emergent species. Species in this group have shown photophobic characteristics, which are rare in young individuals because light is less available in the lower canopy of an intact forest; however, when they have the opportunity, they mostly emerge in the canopy. Group 1 included species that mainly appeared as individuals with a small diameter (<20 cm), such as *Buchanania macrocarpa*, *Endospermum moluccanum* and *Ficus variegata*, which are known as pioneer and fast-growing species. Since they likely appeared equally in smaller diameters, their b-values were close to those of the Group 5 species. The highest b-value observed was 2.25, which was recorded in *Aquilaria polyantha* and *Endospermum moluccanum*. The last two groups (3 and 4) were represented by species with an exponential curve explaining their size class distribution. With a fairly high abundance of smaller individuals, their curves were the short foot of "L-shape". Most species in Group 3 had a maximum DBH of <50 cm, while species in Group 4 had a maximum DBH of >50 cm. However, only a few individuals from both groups were distributed in large diameters, which had little effect on the slope of size class distribution. As shown in Table 6.1, the slope b of these two groups was negative and close to zero. The slope of Group 4 species was close to zero (-0.4 to -3.52, except *Homalium foetidum*) in comparison to that of Group 3 species (-0.4 to -12.05). This result was due to the long tail curve (representing individuals of large diameter). The slope ranges were not strongly differentiated between the two groups since they somewhat overlapped. In Table 6.1, it is also evident that some common species from the two forests were classified interchangeably with Group 3 and Group 4 (e.g., *Pometia acuminata*, *Pometia pinnata*, *Vatica rassak*, *Hopea papuana*, *Anisoptera poly*andra and *Syzygium versteegii*).

Studies of natural populations of tropical tree species with a certain aspect of light illumination are common in most parts of the world, such as Nigeria (Jones, 1950), Surinam (Schulz, 1960), Venezuelan Guiana (Rollet, 1974) and Amazonia (Pires & Prance, 1977).

Distinctions between the size class patterns of shade-tolerant and shade-intolerant species are well studied in tropical America and the Palaeotropics. In the natural rainforest of Surinam, Schulz (1960) noted that the most abundant tree species were represented by a high number of individuals of smaller size class diameters following a normal logarithmic curve pattern. Meanwhile, the size class distributions of a small number of light demanders (e.g., *Goupia glabra*, *Ocotea rubra* and *Quassia amara*) indicated a deficiency in the small and/or middle size classes.

Rollet (1974) recognised seven types of diameter class distribution >10 cm DBH in the Venezuelan forest. He found that species with a very large abundance of individuals of smaller size forming a tall L-shaped curve have an equilibrium population and can regenerate well in moderate shade. On the other hand, he insisted that the early seral trees such as *Cecropia*, *Didymocarpus morototoni* and other light-demanding trees had a size class distribution slightly truncated to the left (with a deficiency in the smaller class), an erratic distribution and an L-shaped curve with a short foot.

Pires & Prance (1977) revealed that in the mature forest of the Jari River, Amazonia, shade-tolerant species (e.g., *Eschweilera* spp. and *Protium tenuifolium*) had a high number of young individuals and mainly exhibited a low or medium stature (C or B storeys) when fully grown. In contrast, young individuals of light-demanding species—such as the remarkably dominant *Bertholletia excelsa*, which mostly presents as an emergent species (50–60 cm in height)—were scarce or lacking. Extremely large emergent species of *Dinizia exelsa* were also found with a similar deficiency in smaller sizes. Other light-demanding species with less extreme emergence (e.g., *Geissorpermum sericeum*) were also found with no smaller diameters in the mature forest.

Although situated far apart on different continents, Jones (1955) and Richards (1996) reported that the size class distribution of shade-tolerant and shade-intolerant species in the Okomu River Nigeria, an old secondary forest, had quite a similar pattern to those in tropical America. He found that species in the B and C storeys dominated the tree population, had numerous young individuals of all sizes and showed a normal logarithmic size class distribution. On the contrary, he found a contradiction among canopy species in the A storey, which formed three types of distribution patterns. First, species with young individuals were hardly investigated in the lower canopy (e.g., *Alstonia boonei* and *Canarium schweinfurthii*). They were classified as strong light-demanding species because they can likely only establish in large canopy gaps. Second, species that were fairly abundant as seedlings and saplings were scarce in the middle size class (e.g., *Guarea thompsonii* and *Lophira alata*). They were grouped as less intolerant species based on Jones' notion that a deficiency in the middle class was only intermittent and that saplings were waiting for a small gap to grow up to the middle class. Third, some species had a normal size class distribution similar to those in the B and C storeys (e.g., *Guarea cedrata* and *Piptadeniastrum africanum*). These were the most shade-tolerant species of the A storey and seemed unable to reach mature size without the help of small gaps.

In Malaysia, Richards (1996) reported the size class distribution of a mixed dipterocarp forest. He noted that the size distribution pattern of dipterocarps was reversed J-shaped with an abundance of seedlings, poles and mature trees but a relatively small number of saplings. Through the variation of the size distribution curve, he revealed that some dipterocarp (e.g., *Shorea maxwelliana*) were more shade tolerant than others (e.g., *Shorea leprosula*). Nevertheless, it was generally emphasised that dipterocarps ranged between very shade tolerant and very light-demanding when compared to large tropical trees on other continents. On the other hand, several large non-dipterocarp trees, which were partly composed of the upper storeys, were somewhat light-demanding. While no emergent trees were found like in America and Africa, the most extreme light demanders were early seral species (e.g., *Macaranga* and *Mallotus* spp.), which appeared transiently in gaps.

Based on the information provided in the aforementioned studies, it can be inferred that there are varying degrees of response to light illumination among tree species, which involve species reflecting their regeneration strategy until reaching the mature phase. Some species are strikingly polarised as either mostly shade-tolerant or mostly shade-intolerant, whereas others are distributed more moderately in this regard based on their level of light tolerance. However, all of these mentioned studies were limited to qualitative interpretation. It is because the shape of nonliniear curve of species' diameter distribution does not clearly differentiate species based on their response to light due to missing individuals in certain size classes and inclination angle of many individuals in small size classes (Brun, 1969 after Mitlöhner, 1998).

Brun (1969) was the first to linearise the tree's stem size distribution such that the whole expression can be reduced to a simple number of slope b from the equation of linear regression (detailed explanation in the previous part of this section). He stated that the more negative b value indicates high shade-tolerant species, and the b value near zero or positive more likely represents light-demanding or pioneer species.

In this regard, the first is the most shade-tolerant species, which presents a normal logarithmic curve, tall L-shaped curve or inverted J-shaped curve that indicates a high abundance of young individuals. These species also mainly form the majority of the tree population and are more likely to have a mature phase in the middle size class or a greater abundance in the B or C storeys than in the A storey. Concerning the Tunas and Bonggo forests, these figures fit the species in Group 2 (Table 6.1a and 6.1b). The second is strong light-demanding species that exhibit erratic size class distributions. While individuals in the smallest size class are rare or lacking, they are most dominant in the large diameter size class (or in the A storey) and even represent emergent species in tropical American and African forests. Other strong light demanders are only found when they regenerate and mature in gaps (e.g., seral species). These descriptions explain most light-demanding species within Groups 1 and 5 in the Tunas and Bonggo forests (Table 6.1a and 6.1b). The third includes the intermediate species as well as less tolerant and intolerant species (i.e., opportunistic species). The size class distribution of these species is indicated by a short L-shaped curve since they are not only found to be very abundant in smaller diameter classes but also have large diameter classes. Typical examples of such species include *Guarea thompsonii* and *Lophira alata* in the Okomu River area, Nigeria (Jones, 1955). Most dipterocarps showed an absence of saplings in a Malaysian forest and were classified as intermediate species (Richards, 1996), which supports the known characteristics of this species group. Accordingly, the distribution patterns of species in Groups 3 and 4 recorded in the Tunas and Bonggo forests (Table 6.1a and 6.1b) were close to the classic distribution pattern for opportunistic species. However, in some cases, certain species can be vaguely interpreted; for instance, the strong shade-tolerant species *Gilbertiodendron dewevrei* (an emergent and dominant species in the single-dominant forest of the eastern Congo Basin) and more shade-tolerant species found among dipterocarps in Malaysian forests that were absent in sampling (Richards, 1996) are unusual if the conditions and history of the forest are unavailable. Regarding this case, the size class distribution of *Vatica rassak* (in Tunas forest) and *Homalium foetidum* (in Bonggo forest) can be difficult to interpret since they can be either more shade-tolerant or opportunistic. Similarly, obscure size distributions were also shown by *Hopea papuana*, *Eucalyptus deglupta*, *Dracontomelum edule* and *Chisocheton ceramicus* (in Bonggo forest). Therefore, additional information from studied forests should be clearly explained or more variable (besides diameter class distribution) and considered when classifying species into light-tolerant types based on stable nitrogen isotopes and total nitrogen content in the xylem.

Furthermore, the linearised curve of size class distribution or slope (b) value determined for each species using the Brun method (Brun, 1969) in Tables 6.1a and 6.1b resulted from the simplification method for quantifying the qualitative value of the non-linear curve. The values strongly differentiated the most shade-tolerant (Group 2) and most light-demanding (Group 5 and 1) species. The b-values of the two extremely tolerant species showed a gradual reduction to zero. This could indicate a decrease in light response from both extremes to a more moderate response, which relates to opportunistic species (Groups 3 and 4).

The b-values of most shade-tolerant species ranged from -67.80 to -17.80 in Tunas forest and from -42.35 to -13.85 in Bonggo forest, whereas those of the most light-demanding species ranged from 0.16 to 2.25 in Tunas forest and between 0.56 and 1.50 in Bonggo forest. The b-values of opportunistic species were between -12.05 and -0.20 in Tunas forest and between -0.40 and -895 in Bonggo forest. These b-values can be used to determine species that are tolerant to light if a forest background is supported and/or additional variables are considered.

6.3. Stable N isotope ratio (δ^{15}N) and total N in tree tissue

The δ^{15}N abundance and total N observed in the sapwood tissue of 103 tree species in the Tunas and Bonggo forests are presented in Table 6.2. The δ^{15}N of 50 species in Tunas forest ranged from -3.900 to 3.785 atom ‰ with a mean of -0.758 ± 1.817 atom ‰, while that of 53 species in Bonggo forest ranged from -2.882 to 4.889 atom ‰ with a mean of 0.320 atom ‰.

Table 6.2.   δ^{15}N abundance (atom ‰) and total N (%) in the sapwood tissue of various tree species classified into five groups based on recruitment and establishment ability.

Group	n	δ^{15}N abundance (atom ‰) (relative to atmospheric N$_2$)			Total N (%)		
		Mean±SD	Min	Max	Mean±SD	Min	Max
				Tunas forest			
All	50	-0.758 ± 1.817	-3.900	3.785	0.192 ± 0.083	0.110	0.533
1	3	2.562[a] ± 1.415	1.013	3.785	0.394[a] ± 0.147	0.240	0.533
2	9	-2.704[bc] ± 0.709	-3.900	-2.172	0.138[b] ± 0.014	0.120	0.160
3	19	-1.496[cd] ± 1.312	-3.568	0.865	0.167[ab] ± 0.040	0.110	0.240
4	7	-0.126[d] ± 0.799	-1.550	0.823	0.167[ab] ± 0.055	0.110	0.240
5	12	0.510[ad] ± 1.362	-1.564	3.074	0.236[a] ± 0.070	0.150	0.340
				Bonggo forest			
All	53	0.320 ± 1.778	-2.882	4.889	0.192 ± 0.082	0.090	0.430
1	7	2.166[a] ± 1.505	0.429	4.889	0.293[a] ± 0.110	0.150	0.430
2	5	-1.672[b] ± 0.685	-2.852	-1.080	0.114[b] ± 0.011	0.100	0.130
3	12	-0.368[bc] ± 1.473	-2.882	2.823	0.150[b] ± 0.052	0.090	0.240
4	12	-0.037[bc] ± 1.596	-2.183	2.397	0.163[ab] ± 0.029	0.110	0.220
5	17	0.884[ac] ± 1.595	-1.161	4.512	0.224[a] ± 0.074	0.150	0.390

The $\delta^{15}N$ abundance reduced gradually from the most light-demanding species (Group 1) to the most light-intolerant species (Group 2). Group 1 had a $\delta^{15}N$ abundance range slightly above zero (1.013–3.786 atom ‰ and 0.429–4.889 atom ‰ in Tunas and Bonggo, respectively) with the highest recorded means of 2.562 atom ‰ in Tunas forest and 2.166 atom ‰ in Bonggo forest. These values differed significantly from those of Group 2 (-2.704 atom ‰ in Tunas and -1.672 atom ‰ in Bonggo forest), which ranged far below zero (-3.900 to -2.172 atom ‰ and -2.852 to -1.080 atom ‰). Group 5 $\delta^{15}N$ abundance showed no significant difference when compared to Group 1, demonstrating a typical result for most light-demanding species. Moreover, some species values ranged slightly below zero (negative), which is unusual for Group 1 species. The intermediate species (Groups 3 and 4) showed no significant difference in $\delta^{15}N$ abundance when compared to the most light-demanding and light-intolerant groups. However, they exhibited transition values between the two species with extreme light requirements.

The total nitrogen exhibited by the sapwood tissue of the 102 tree species varied between 0.110% and 0.533% in Tunas forest and between 0.090% and 0.430% in Bonggo forest, with an equal mean of 0.192% ± 0.08% in both forests. The variations in total N content showed a positive correlation with ^{15}N abundance among groups of species, with the most light-demanding species showing the highest N content and gradually declining to the lowest N content in the most light-intolerant species. Although the correlation between the two N values contained in sapwood tissue revealed a weak relationship (r^2 = 0.536 and 0.519 for the Tunas and Bonggo forests, respectively), the linear regression confirmed that positive $\delta^{15}N$ is more likely to be associated with higher N content (>0.22%) in sapwood tissue (Figure 6.1). Additionally, the constant slope "a" and slope "b" of the linear regression line explained that the $\delta^{15}N$-N curve of Tunas forest was below that of Bonggo forest, with a narrower difference at higher N than lower N.

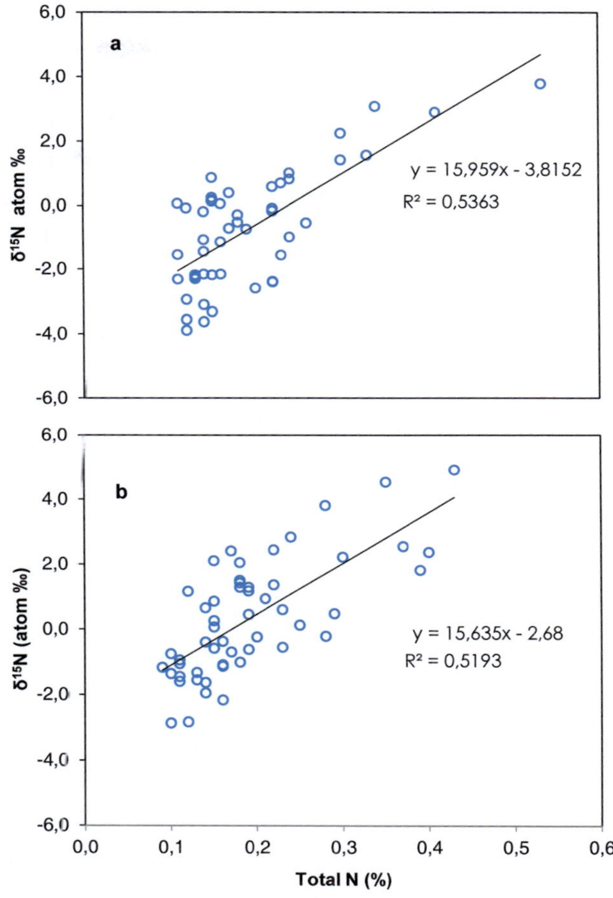

Figure 6.1. Correlation of δ15N abundance (atom ‰) and total N (%) in the sapwood tissue of 50 tree species from Tunas forest (a) and 53 tree species from Bonggo forest (b).

Comparing the median of all species between two forests, a Mann-Whitney U-test (Table 6.3) indicated that the species in Bonggo showed significantly higher $\delta^{15}N$ abundance than those in Tunas (p<0.01); however, no significant difference was observed in their N content (p>0.05). Groups from the two forests only showed significantly different $\delta^{15}N$ abundance and N content between species in Group 2 (p<0.05) but not in other groups. The total N contained in the sapwood of species in Group 2 was ranked higher in Tunas forest than in the same group of species in Bonggo forest; on the contrary, ^{15}N abundance in the same species group was ranked in inverse order for the two different forests.

Table 6.3. Median test for $\delta^{15}N$ abundance (atom ‰) and total N (%) in the sapwood tissue of five groups of tree species in the Tunas and Bonggo forests.

Group	$\delta^{15}N$ abundance (atom ‰) (relative to atmospheric N_2)					Total N (%)				
	n_T	n_B	Med_T	Med_B	p-value	n_T	n_B	Med_T	Med_B	p-value
All	50	53	-0.716	0.101	0.0053[**]	50	53	0.160	0.170	0.9973[ns]
1	3	7	2.888	2.344	0.5166[ns]	3	7	0.410	0.290	0.2667[ns]
2	9	5	-2.299	-1.471	0.0295[*]	9	5	0.140	0.110	0.0119[*]
3	19	12	-1.447	-0.875	0.0926[ns]	19	12	0.150	0.135	0.2198[ns]
4	7	12	-0.084	-0.089	0.9671[ns]	7	12	0.180	0.165	0.9018[ns]
5	12	17	0.273	0.839	0.8446[ns]	12	17	0.225	0.190	0.6160[ns]

n = number of sample test; Med= median; ** = p<0.01; * p<0.05; [ns]= not significant
Subscript $_T$ = Tunas forest; $_B$ = Bonggo forest

The $\delta^{15}N$ abundances and N content reported here are similar to the findings of others (Hoering, 1955; Pakwel *et al.*, 1957; Virginia & Delwiche, 1982). As shown in Table 6.4, the results vary between species and the parts of trees sampled. These variations may result from several factors, including the organ-specific loss of nitrogen, different patterns of nitrogen assimilation and the reallocation of nitrogen (Pate, 1973; Evans, 2001). With the present data, it is difficult to quantitatively assess which of these has produced the observed nitrogen and its isotope variations.

Table 6.4. $\delta^{15}N$ abundance (atom ‰) and total N (%) in various parts of trees from subtropical and tropical forests.

Name	Part of the sampled tree	$\delta^{15}N$ (‰)	N(%)	Sources
White clover	Foliar	-6.5		Hoering (1955)
Dandelion	Foliar	-2.8		Hoering (1955)
Red oak	Foliar	-0.9		Hoering (1955)
Cedar	Foliar	+1.3		Hoering (1955)
American elm	Foliar	+1.9		Hoering (1955)
Picea abies	Wood	-3.8	0.04	Pakwel *et al.* (1957)
Betula alba	Wood	-2.2	0.12	Pakwel *et al.* (1957)
Betula alba	Outer bark	-5.0	0.62	Pakwel *et al.* (1957)
Alnus oregona	Foliar	+0.1		Virginia & Delwiche (1982)
Pinus muricata	Foliar	+1.6		Virginia & Delwiche (1982)
Quercus kelloggii	Foliar	+0.4		Virginia & Delwiche (1982)
Pinus sabiniana	Foliar	+0.8		Virginia & Delwiche (1982)
Pinus albicaulis	Foliar	-0.3		Virginia & Delwiche (1982)
Abies concolor	Foliar	-0.8		Virginia & Delwiche (1982)
Prunus emarginata	Foliar	-0.5		Virginia & Delwiche (1982)
Calocedrus decurrens	Foliar	-1.0		Virginia & Delwiche (1982)
Tsuga canadensis	Xylem	+0.9	0.12	Poulson *et al.* (1995)
Tsuga canadensis	Phloem	+0.4	0.34	Poulson *et al.* (1995)
Tsuga canadensis	Outer bark	+1.7	0.22	Poulson *et al.* (1995)
Acacia mangium	Xylem	+1.5	0.21	Mitlöhner, unpublished
Bing	Xylem	+2.1	0.31	Mitlöhner, unpublished
Cecropia sp.	Xylem	+2.6	0.35	Mitlöhner, unpublished
Chrodo	Xylem	-0.52	0.23	Mitlöhner, unpublished
Eucalyptus sp.	Xylem	+3.6	0.21	Mitlöhner, unpublished

Despite the limitations of the information obtained from this study, the pattern of nitrogen assimilation among species can be determined from previous studies. A laboratory study of bleeding xylem sap collected from a range of herbaceous plants that were continuously supplied with culture solution (140 parts/10^6 NO$_3$–N) indicated a range of combinations between free nitrate and organic nitrogen in xylem fluids (Pate, 1973). This configuration was claimed as a useful indicator of the ability of various woody plant species' roots to assimilate incoming nitrate.

Nitrogen uptake in the root system is mainly in the forms of NH_4^+ and NO_3 (Ullrich, 1992). The assimilation of ammonium occurs only in the root system by the glutamine synthetase–glutamate synthase (GS–GOGAT) pathway, whereas nitrate assimilation can occur in roots, leaves, or both via the nitrate reductase–nitrite reductase pathway (Evans, 2001). Nitrogen sources and patterns of assimilation can strongly influence the composition of $\delta^{15}N$ in different parts of plants (Yoneyama *et al.*, 1991; Evans *et al.*, 1996). Since the assimilation of NH_4^+ occurs immediately in roots (Bloom, 1988), minor variations in ^{15}N abundance among the parts of a plant can be observed when NH_4^+ is the main source of nitrogen. On the contrary, significant variation arises when NO_3^- is the primary nitrogen source. The bleeding xylem sap study exhibited an extreme situation in which nitrate hardly presented in the xylem, while a high level of organic solutes of nitrogen existed when the reductase enzyme completely assimilated nitrate in the roots. These outermost species have their shoots ready to synthesise organic nitrogen. In another extreme situation, plants in which nitrate reductase was not detected in the root system and xylem regularly contained 95–99% free nitrate. When ammonium or urea was added to the external medium, the plants demonstrated that amides and amino acids could be readily synthesised and exported from the root system.

Nitrate reductase and glutamine synthetase both fractionate against ^{15}N, while the observed discrimination by nitrate reductase and glutamine synthetase is 15 and 17‰, respectively (Handley & Raven, 1992; Yoneyama *et al.*, 1993). This fractionation causes N in root tissue to become enriched in ^{15}N when compared to the N outside. The organic nitrogen is then depleted in ^{15}N, while the ^{15}N of unassimilated NO_3^- becomes more enriched relative to organic N (Evans, 2001).

Concerning the ^{15}N abundance presented here, the tree species of Group 2 were most likely to experience high nitrate fractionation in their root systems. The excess depleted ^{15}N (organic nitrogen), which was exported to the stem and shoots, reflected a very low ^{15}N abundance. On the other hand, the tree species of Groups 1 and 5 were presumed to have extremely low root nitrate reductase activity. A high abundance of unassimilated nitrates exported from the roots to the stems and then to the shoots was detected by a high abundance of $\delta^{15}N$ in sapwood tissue. Thus, reductase enzyme activity gradually increased within the intermediate species from Group 4 to Group 3.

These results are in line with the expectation that light-demanding and shade-tolerant species have higher and lower $\delta^{15}N$ abundances, respectively. This notion was supported by a xylem sap study (Pate, 1973), indicating that root reduction of nitrate occurs extremely slowly and results in very low levels of free amino compounds in **light-demanding species**. Pate (1973) suggested that it is reasonable to link nitrate reductase enzymes to the photosynthetic process facilitating the amino acid requirements for protein synthesis. Indeed, classic nitrate-storing plants (e.g., perennial herbaceous plants such as *Borago*, *Beta*, *Chenopodium* and *Menyanthes*) in which, the levels of nitrate in leaves or stems may represent up to 5–10% of the plant dry weight (McKee, 1962 after Pate, 1973) serve as examples of the nitrogen content of light-demanding species. In contrast, studying the pattern of nitrate reductase distribution has indicated that species with high root reductase activity that saves excess nitrogen in their shoot systems tend to store the same family of compounds in their stems and leaves. Examples of this phenomenon include allantoin storage in *Phaseolus* shoots, asparagine storage in *Trifolium* and *Lupinus*, and large reserves of citrulline in alder (*Alnus*) and of alkaloids in tobacco (*Nicotiana*). A similar strategy can also be found in some *Canarium* species, which have a particular aromatic smell that indicates organic nitrogen storage. Moreover, *Canarium* species were assumed to be the most shade-tolerant species included in this study. Common compounds present in the xylem include the amides glutamine and asparagine as well as other amino acids (Pate, 1973). Notably, these compounds have an N: C ratio of 0.4 (Bollard, 1960 after Pate, 1973), indicating low nitrogen content in comparison with free nitrate.

The observed resource limitations in Tunas forest when compared to Bonggo forest (e.g., lower soil nitrogen content and lower precipitation) (see Chapter 2) are presumed to cause the higher median ^{15}N and lower N values in the sapwood tissue of species in Group 2. It has been argued that more stress-tolerant plants retain greater amounts of nitrogen, while lower $\delta^{15}N$ values are caused by the efflux of nitrogen compounds with positive $\delta^{15}N$ values (Robinson *et al.*, 2000).

6.4. Diameter distribution, stable N isotope ratio ($\delta^{15}N$) and total N

Tree species size classes had a significant relationship with $\delta^{15}N$ abundance and nitrogen content in sapwood tissue ($p<0.05$). Figure 6.2 shows an exponential relationship between b-value and $\delta^{15}N$ abundance with a correlation coefficient (R) of 0.669 for tree species in Tunas forest and 0.616 for those in Bonggo forest. The graphs explain that the isotope abundance ratio decreased dramatically from approximately 4.0 and 5.0 atom ‰ to -1.6 and -1.1 atom ‰ as the positive b-value moved from approximately 2.5 and 2.0 to 0 and the abundance began to gradually reduce to asymptotes of -2.067 and -1.589 atom ‰ as the b-value increased in a negative value for tree species in the Tunas and Bonggo forests, respectively. In addition to nitrogen content, Figure 6.3 also displays the exponential relationship of total nitrogen content in sapwood tissue via species b-values with correlation coefficients of 0.839 and 0.755 for tree species in the Tunas and Bonggo forests, respectively. The graphs indicate sharp nitrogen reductions of approximately 4.0% and 3.2% when positive b-values approached zero from their maximum values and gradual 0.145% and 0.131% decreases in values for tree species in the Tunas and Bonggo forests, respectively. Thereafter, the N content remained constant despite the b-values of species increasing.

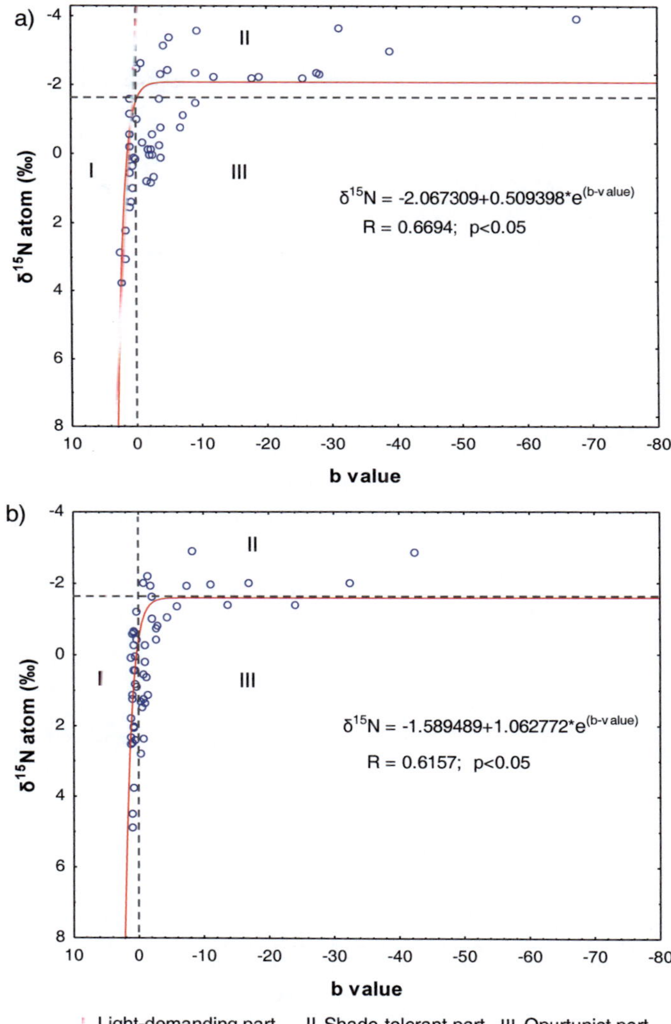

Figure 6.2. Correlation between the b-value and $\delta^{15}N$ atom (%) of 50 tree species from Tunas forest (a) and 53 tree species from Bonggo forest (b).

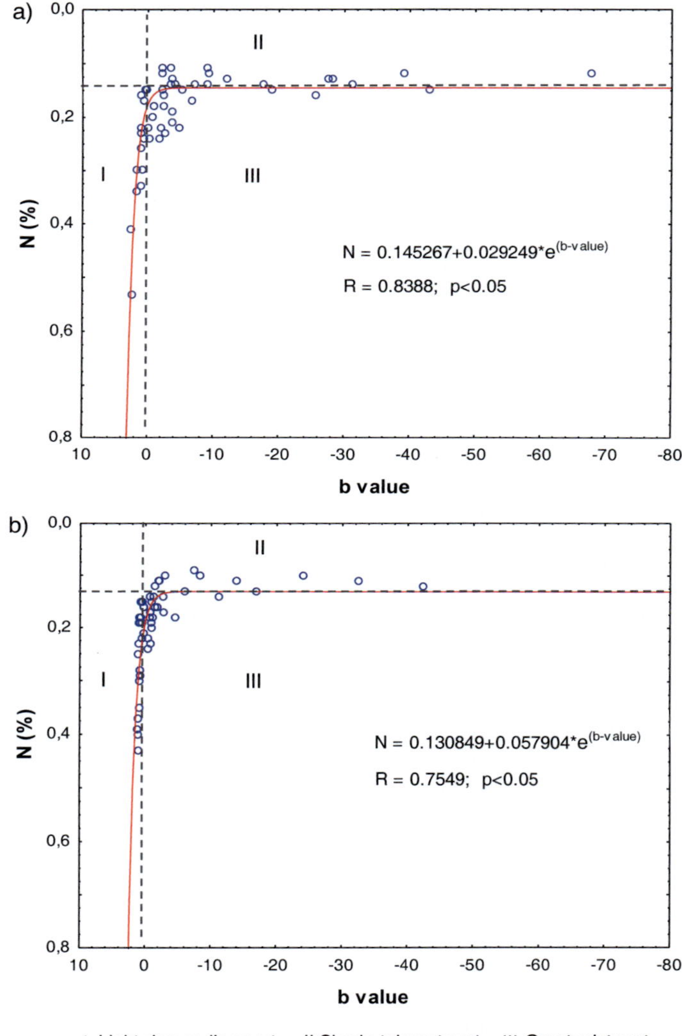

Figure 6.3. Correlation between the b-value and total nitrogen (%) of 50 tree species from Tunas forest (a) and 53 tree species from Bonggo forest (b).

The curves that related species' diameter size distribution and species' nitrogen content were differentiated into three parts. The first part was the asymptotic portion. Species mostly formed the pattern of the curve with negative b-values and their nitrogen indicators ($\delta^{15}N$ and total N values), which were less than or equal to the asymptotic line. Notably, there was no significant correlation between variables (b-values and nitrogen indicators) in this part. The second part was the gradual portion. The pattern of this part was mostly created by negative b-values and nitrogen values greater than the asymptote. In this part, the negative b-value exhibited a significant correlation with the nitrogen indicators. The third part was the sharp curves. In this part, only the positive b-value contributed enormously to the nitrogen indicators as it increased far from zero.

As previously mentioned, in order to serve a role in the variation of nitrogen content in xylem sap, nitrate reductase is presumed to influence the plant's physiological mechanisms, which allowed the grouping of tree species based on three N-slope curve patterns. The asymptotic curve group comprises species with high nitrate reductase activity in the root system. These species are characterised by a high depleted isotope ratio product and an increased ability to store organic nitrogen in root and stem tissue (Evans, 2001). Such characteristics are predicted to allow these species to have a range of nitrogen indicators, from -1.5 to -4.8 atom ‰ and from 0.09 to 0.15% for ^{15}N and N content, respectively. These also are predicted to be nutrient-conservative species that can grow abundantly with limited light illumination below a forest canopy. The sharp curve group included species assumed to have high shoot system reductase activity, which is linked to the photosynthetic process. These species transported high amounts of enriched nitrate and small amounts of the free amino compound, as indicated by the high ^{15}N abundance and nitrogen content in the xylem sap to be synthesised in the shoot system. These species favour canopy gaps to start their growth and competition to reach the upper canopy. The gradual curve group included species predicted to have nitrate assimilation in both root and shoot systems. The intensity of the reductase enzyme is predicted to vary between the two assimilation sites. If reductase enzyme activity is high in the shoot system, then the nitrogen indicator and size class distribution will be similar for the sharp curve species. Notably, the inverse relationship occurs if the intensity is high in the root system. A gradual increase in the value of variables is typical of these intermediate species.

6.5. Grouping of tree species light requirements

The exponential relationships observed from plotting species b-values against $\delta^{15}N$ or total N content (see Section 6.2.) successfully outlined species-specific light requirements based on 50 tree species from Tunas forest and 53 species from Bonggo forest. Grouping species into three categories (i.e., light-demanding, opportunistic and shade-tolerant) was based on their position in the three parts of the exponential curve. A vertical line crossing b-value = 0 indicated the point of distinction between light demanders (positive values) and shade-tolerant and opportunistic species (negative values). A horizontal line on the average asymptote for each curve ($\delta^{15}N$=-1.6 atom ‰ and total N=0.14) differentiated shade-tolerant species (smaller values) and opportunists (larger values). Accordingly, species with steeper slopes of the curves ranged from moderate to extreme light demanders, which had b-values >0, $\delta^{15}N$ >-1.6 atom ‰ and total N >0.14%. The greater the three indicator values, the higher the light requirement. The next group was species distributed along the asymptote at negative b-values, with $\delta^{15}N$ <-1.6 atom ‰ and total N <0.14%. Notably, these were categorised as shade-tolerant species. While species in this group were mostly indicated by a narrow range of the lowest N indicators, their wide variation in b-value characterised their ability to exist under limited light conditions. Species with negative b-values far from zero represented extremely shade-tolerant species. The final group included the opportunistic species. These were more concentrated at the corner of the vertical and horizontal lines. The light intensity requirements of species from this group were primarily reflected by their N indicator values rather than their b-values. The higher the N indicator values, the more the species responded to light than shading.

Out of the 50 tree species from Tunas forest, 15 were placed into the light-demanding group, 15 were placed into the shade-tolerant group and 20 were categorised as opportunists (Table 6.5a). On the other hand, out of the 53 tree species from Bonggo forest, 25 were considered light demanders, 14 were considered shade tolerant and the remaining 14 were considered opportunists (Table 6.5b).

Table 6.5a. Light requirement characteristics of 50 tree species from **Tunas** forest based on the slope of the diameter distribution (b-value), depletion isotope $\delta^{15}N$ and total N.

Group	Slope b	$\delta^{15}N$ (atom ‰)	N (%)	Species name	Characters
Light-demanding	2.50	+2.889	0.41	*Aquilaria polyantha*	High
	1.47	+2.243	0.30	*Aglaia* sp.	High
	0.60	-0.179	0.22	*Alstonia spectabilis*	Moderate
	0.38	+0.581	0.22	*Anisoptera thurifera*	Moderate
	0.50	-0.542	0.26	*Campnosperma auriculatum*	Moderate
	0.49	+0.386	0.17	*Cinnamomum culilawan*	Moderate
	0.25	+0.136	0.15	*Dillinea ovalifolia*	Moderate
	2.25	+3.785	0.53	*Endospermum moluccanum*	High
	1.60	+3.074	0.34	*Gironniera subaequalis*	High
	0.50	+1.013	0.24	*Gnetum gnemon*	High
	0.61	+1.411	0.30	*Neoscortechinia* sp.	High
	0.91	+1.561	0.33	*Pimelodendron amboinicum*	High
	0.14	-1.564	0.23	*Podocarpus nerlifolius*	Moderate
	0.02	-1.146	0.16	*Syzygium gustavioides*	Moderate
	0.16	+0.161	0.15	*Xanthostemon crenulatus*	Moderate

Table 6.5a. Continued

Group	Slope b	$\delta^{15}N$ (atom ‰)	N (%)	Species name	Characters
Shade-tolerant	-18.90	-2.192	0.15	*Beilschmiedia* sp.	High
	-39.10	-2.959	0.12	*Canarium asperum*	High
	-31.25	-3.636	0.14	*Canarium hirsutum*	High
	-25.70	-2.173	0.16	*Canarium indicum*	High
	-19.75	-2.172	0.14	*Garcinia dulcis*	High
	-43.10	-2.248	0.15	*Garcinia celebica*	High
	-5.15	-3.332	0.15	*Litsea ledermannii**	Moderate
	-67.80	-3.900	0.12	*Litsea timoriana*	High
	-28.25	-2.286	0.13	*Palaquium lobbianum*	High
	-4.25	-3.122	0.14	*Palaquium obtusifolium*	Moderate
	-5.90	-2.325	0.11	*Planchonella firma*	High
	-12.05	-2.196	0.13	*Pometia acuminata*	High
	-3.28	-3.568	0.12	*Pometia pinnata**	Moderate
	-27.70	-2.313	0.13	*Syzygium anomalum*	High
	-3.75	-2.262	0.13	*Vitex pubescens*	Moderate
Opportunist	-3.75	0.158	0.21	*Aceracium diversivolium*	Light+
	-1.04	-0.295	0.18	*Anisoptera polyandra***	Light+
	-4.80	-2.397	0.22	*Blumeodendron amboinicum*	Unclear
	-2.15	+0.057	0.11	*Calophyllum inophyllum**	Shade+
	-13.52	-1.550	0.11	*Calophyllum papuanum**	Shade+
	-0.20	-2.419	0.22	*Chisocheton ceramicus*	Unclear
	-2.55	-0.527	0.18	*Hopea papuana***	Light+
	-3.50	-0.196	0.14	*Lithocarpus rufovillosus*	Shade+
	-11.75	-0.716	0.17	*Litsea tuberculata*	Shade+
	-7.29	-1.080	0.14	*Litsea irianensis*	Shade+
	-2.30	+0.865	0.15	*Mastixiodendron plectocarpum*	Light+
	-2.45	+0.052	0.16	*Myristica globosa*	Light+
	-0.25	-0.988	0.24	*Myristica hollrungii*	Light+
	-0.75	-2.617	0.20	*Myristica lepidota*	Unclear
	-3.75	-0.741	0.19	*Podocarpus amara*	Light+
	-1.96	-0.091	0.22	*Spathiostemon javensis*	Light+
	-2.24	-0.084	0.12	*Syzygium hylophilum**	Shade+
	-1.74	+0.823	0.24	*Syzygium versteegii*	Light+
	-2.63	+0.694	0.23	*Vatica rassak*	Light+
	-13.77	-1.447	0.14	*Whiteodendron papuana*	Shade+

Aquilaria polyantha and *Endospermum moluccanum* were classified as the most light-demanding species in the Tunas forest. This was indicated by the greatest positive b-values (2.5 and 2.25, respectively), which were in relation to the high positive isotope $\delta^{15}N$ of +2.889 and +3.785 atom ‰ as well as high total N content of 0.41% and 0.53%, respectively. In the undisturbed Tunas forest, these two species were rarely found and had a maximum DBH of 20 cm and a maximum height of <12 m. On the other hand, *Podocarpus nerlifolius*, *Syzygium gustavioides* and *Campnosperma auriculatum* were considered moderate light demanders. Despite having the lowest positive b-values, these species also had $\delta^{15}N$ values that were near the asymptote of -1.6 atom ‰ as well as considerable N content. In Tunas forest, these species had a low abundance of smaller-sized trees and one or two large-sized trees in the upper canopy (b-value <1).

While all species in the shade-tolerant group exhibited very high $\delta^{15}N$ depletion, their abundance of small-sized trees varied. *Litsea timoriana*, *Garcinia celebica* and *Canarium asperum* were selected as the most shade tolerant since they had the highest negative b-values of -67.80, -43.10 and -39.10, which were correlated with high $\delta^{15}N$ depletion values of -3.900, -2.248 and -2.959 atom ‰, respectively. *Garcinia celebica* and some other species had a total N over the limit of shade tolerance (0.14%). *Litsea timoriana*, *Garcinia celebica* and *Canarium asperum* had a high abundance of individuals in the middle and lower canopies; thus, they were listed as the most dominant species in the Tunas forest. On the contrary, *Palaquium obtusifolium*, *Vitex pubescens*, *Pometia pinnata* and *Litsea ledermannii* were recorded as moderate shade-tolerant species due to their low negative b-values. In Tunas forest, the abundances of moderate shade-tolerant species in the lower canopy were lower than those of high shade-tolerant species.

It could also be determined whether species that tolerate both light and shade conditions (i.e. opportunists) preferred light more than shading (light+), or the opposite preference (shade+). *Vatica rassak* accounted for a considerable number of small-sized trees (b-value = -2,63) and exhibited a weak ability to assimilate nitrogen ($\delta^{15}N$ = +0,694 atom ‰) and a high level of unassimilated nitrogen (N = 0.23%). This species was classified as an opportunist that would grow better in light conditions (light+); however, it can also survive in shaded areas. *Calophyllum papuanum* accounted for a considerable number of small-sized trees (b-value = -13.52), a high ability to assimilate nitrogen ($\delta^{15}N$ =-1,550 atom ‰) and low unassimilated nitrogen content (N = 0.11%). Moreover, it was considered an opportunist that requires shade (shade+). Species such as *Blumeodendron amboinicum*, *Chisocheton ceramicus* and *Myristica lepidota* were considered unclear because they had a strong ability to reduce $\delta^{15}N$ abundance ($\delta15N$ > -2.000 atom ‰); however, they simultaneously had high unassimilated nitrogen content in sapwood (N >0.20%). *Blumeodendron amboinicum* belongs to the family *Euphorbiaceae*, which also had species (e.g., *Neoscortechinia* sp.) listed as light+. *Chisocheton ceramicus* was recorded in the Bonggo forest as a light+ opportunistic species. In the case of *Myristica lepidota*, all other *Myristica* species in Tunas forest were assigned as having a stronger light preference (light+).

Table 6.5b. Light requirement characteristics of 53 tree species from **Bonggo** forest based on the slope of the diameter distribution (b-value), depletion isotope $\delta^{15}N$ and total N.

Group	Slope b	$\delta^{15}N$ (atom ‰)	N (%)	Species name	Characters
	0.820	-0.637	0.190	*Albizia falcataria*	Moderate
	0.700	+2.086	0.160	*Alphitonia macrocarpa*	High
	0.820	+3.789	0.280	*Alstonia scholaris*	High
	0.107	+0.101	0.250	*Anisoptera thurifera*	Moderate
	1.250	+1.789	0.390	*Anthocephalus cadamba*	High
	1.500	+2.344	0.400	*Aquilaria* sp.	High
	1.500	+2.529	0.370	*Buchanania macrocarpa*	High
	0.175	+1.265	0.145	*Campnosperma auriculatum*	Moderate
	0.750	+2.025	0.180	*Cananga odorata*	High
	0.447	+0.044	0.150	*Cinnamomum culilawan*	Moderate
Light-demanding	0.558	+0.840	0.150	*Dillenia alata*	Moderate
	0.570	+4.512	0.350	*Duabanga moluccana*	High
	0.790	-0.236	0.280	*Endospermum medullosum*	Moderate
	0.371	+2.183	0.160	*Eucalyptus deglupta*	High
	0.500	+0.429	0.190	*Ficus variegata*	Moderate
	0.750	+0.457	0.290	*Gnetum gnemon*	Moderate
	0.970	+1.149	0.190	*Intsia bijuga*	Moderate
	1.000	+4.889	0.430	*Linociera oblongifolia*	High
	0.870	+2.203	0.300	*Macaranga beillei*	High
	0.660	-0.608	0.150	*Parartocarpus* spp	Moderate
	0.475	-1.161	0.160	*Pimelodendron amboinicum*	Moderate
	0.250	-0.398	0.160	*Podocarpus nerlifolius*	Moderate
	0.990	-0.568	0.230	*Pterocarpus indicus*	Moderate
	0.250	+0.919	0.210	*Spondias dulcis*	Moderate
	0.500	+2.429	0.220	*Vitex* sp.	High

Table 6.5b. Continued

Group	Slope b	δ¹⁵N (atom ‰)	N (%)	Species name	Characters
Shade-tolerant	-24.050	-1.380	0.110	*Calophyllum inophyllum**	Moderate
	-13.850	-1.080	0.110	*Calophyllum papuanum**	Moderate
	-42.350	-2.852	0.120	*Canarium asperum*	High
	-16.900	-1.577	0.130	*Canarium indicum*	Moderate
	-6.750	-1.646	0.130	*Celtis rigescens*	High
	-7.000	-1.973	0.110	*Dysoxylum amooroides*	High
	-11.180	-1.652	0.140	*Homalium foetidum*	High
	-12.400	-1.971	0.090	*Mangifera* sp.	High
	-8.950	-1.778	0.100	*Mastixiodendron pachyclados*	High
	-0.710	-2.396	0.120	*Palaquium obtusifolium*	High
	-0.700	-1.970	0.130	*Pometia acuminata*	High
	-2.380	-1.617	0.110	*Pterygota horsfieldii*	High
	-32.450	-1.471	0.110	*Syzygium anomalum*	Moderate
	-8.290	-2.882	0.100	*Syzygium hylophilum**	High
Opportunist	-2.680	-0.412	0.150	*Anisoptera polyandra***	Shade+
	-0.880	+1.268	0.190	*Chisocheton ceramicus*	Light+
	-2.430	+1.143	0.140	*Diospyros* sp.	Light+
	-1.080	+1.394	0.180	*Dracontomelum edule*	Light+
	-2.750	-0.724	0.170	*Hopea papuana***	Shade+
	-0.640	+1.483	0.180	*Litsea ledermannii**	Light+
	-0.350	-0.263	0.200	*Myristica lepidota*	Light+
	-3.250	+0.627	0.140	*Myristica* sp.	Unclear
	-0.930	+0.233	0.150	*Pometia pinnata**	Light+
	-4.510	-1.037	0.180	*Syzygium* sp.	Shade+
	-0.400	+2.823	0.240	*Syzygium versteegii*	Light+
	-1.800	-1.116	0.160	*Terminalia arborea*	Shade+
	-0.400	+1.339	0.220	*Toona sureni*	Light+
	-0.800	+0.578	0.230	*Vatica rassak*	Light+

In the Bonggo forest, *Anthocephalus cadamba*, *Aquilaria* sp., *Buchanania macrocarpa*, *Duabanga moluccana* and *Linociera oblongifolia* were considered the most light-demanding based on their high positive b-values with respect to others as well as their high $\delta^{15}N$ and N values. Their b-values ranged from 0.570 to 1.500, their $\delta^{15}N$ values ranged from +1,789 to +4,889 atom ‰, and their total

N values were between 0.350% and 0.430%. These species were mostly found in the middle and lower canopy with heights not exceeding 18 m—except for *Anthocephalus cadamba* and *Duabanga moluccana*, which had one or two trees with a height of less than 25 m. Nevertheless, while moderately light-demanding species such as *Intsia bijuga* were dominant in the upper canopy (even as emergent species), small-sized trees were absent. *Albizia falcataria, Pterocarpus indicus, Cinnamomum culilawan* and others listed as moderate light-demanding species were also found as large-sized trees in the upper canopy and scarcely as small trees in the lower canopy. However, these trees were not as abundant as *Intsia bijuga*.

Although they varied greatly in b-values, species belonging to the shade-tolerant group in the Bonggo forest had a narrower range of $\delta^{15}N$ values. While some species could not even reach the $\delta^{15}N$ value limit (-1.6 atom ‰) for shade-tolerant species, they had highly negative b-values (e.g., *Calophyllum inophyllum, Calophyllum papuanum, Canarium indicum* and *Syzygium anomalum*). Hence, they were listed as moderately shade tolerant. However, the most common species (e.g., *Canarium asperum, Palaquium obtusifolium* and *Syzygium hylophilum*) were the most shade tolerant due to their high ability to reduce nitrate and their low sapwood nitrogen content.

Syzygium sp., *Terminalia arborea, Anisoptera poly*andra and *Hopea papuana* were among the opportunists in the Bonggo forest that most strongly favoured shading over light (shade+) since they had a lower abundance of ^{15}N in their sapwood when compared to the atmosphere. Other species, which were marked as Light+, had a higher abundance of ^{15}N in their sapwood than the standard.

Nearly half (n=22) of the studied species were common in both the Tunas and Bonggo forests. However, not all of the shared species exhibited similar light requirements, which would result in them being classified in the same group. The results indicated that only five unmatched species out of the 22 species were shared between shade-tolerant and opportunistic groups; however, these were not included in the light-demanding group. In one case, *Calophyllum inophyllum*,

Calophyllum papuanum and *Syzygium hylophilum* registered as opportunist shade+ species in Tunas forest and were assigned as moderately shade tolerant in Bonggo forest. Moreover, *Litsea ledermannii* and *Pometia pinnata*—listed as moderately shade tolerant in Tunas forest—were grouped as opportunistic light+ species in Bonggo forest. This result likely stems from the first three species having minimal nitrate assimilation in roots when their resources were lower in the soil of Tunas forest; thus, ^{15}N reduction was low in sapwood (+0,057 to -1,550 atom ‰). In the Bonggo forest, abundant nitrate was likely available such that the species assimilated nitrate more in their roots; thus, a high ^{15}N reduction was detected in sapwood (-1.080 to -2.882 atom ‰). The two remaining species, *Litsea ledermannii* and *Pometia pinnata*, were considered opportunists with typical light+. Although they could assimilate nitrate in their roots as opportunists, they also preferred ammonium to be fixed in leaves. However, the ammonium resources were probably less available in the Tunas forest than in the Bonggo forest; therefore, the species had to adjust their abilities based on the soil conditions. Moreover, the observed shift in species' response to light implied that the nitrogen resources in Tunas forest were not as good as those of Bonggo forest.

Ultimately, the results of the present study are in agreement with the expectation that the variation of $\delta^{15}N$ and N in sapwood can elucidate species' light requirements. The present study successfully exhibited the principal roles of $\delta^{15}N$ and N in differentiating species into three light requirement groups and identifying specific light-responding characteristics (ranging from high light-demanding species to high shade-tolerant species). Relevant studies from the region suggest that this is the first outstanding finding on species' light requirements. Hence, further investigations should be conducted at the greenhouse and field level to confirm the results of this study.

7. Silvicultural approaches and further implications

7.1. Silviculture: General definitions and implications

The Society of American Foresters defines silviculture (after Ford-Robertson, 1971; Helms, 1998) as the skill of managing the production and cultivation of forest stands by implementing scientific knowledge in maintaining the establishment, composition, growth, health and quality of the forest; the application of human-involved maintenance to improve forest productivities and benefits to an owner and society sustainably; and the integration of biologic and economic models that enhance treatment methods to meet the objectives and satisfaction of a stand owner. Moreover, silviculture concerns techniques related to treatment that can improve the artificial stand composition and structure to mimic the intact forest characteristics and to maintain biodiversity and the essential functions and values of the forest ecosystem (Oliver, 1992; Bergeron *et al.*, 1999). This definition is consistent with (Lamprecht, 1989) statement that the original forest stand is the basis of decision-making in a suitable silviculture treatment.

In general, silviculture is designed to provide desired values to the greatest extent possible such that a stand can exist sustainably. It requires a comprehensive view that relates environmental effects to the establishment and development of an individual tree, species composition, diversity and stand structure in a forest community. It is challenging to provide such options that must ecologically ensure appropriate long-term stand management and, at the same time, to have the capability to identify the best choice that serves the interests of the landowner (after Broun, 1912; Nyland, 2016).

Silviculturists must be interested in forest ecology or silvics to achieve better silvicultural techniques that fit the sustainable practices and needs of an owner. Silvics, a particular branch of forest ecology, deals with how individual trees respond to changes in the physical environment when growing, developing and reproducing (Smith, 1986; Helms, 1998). It also investigates how the tree population, in turn, modifies the physical environment as well as how vegetation–environment interaction transforms the forest over a long period (Smith, 1986). However, silviculture is both a science and an art. Accordingly, silviculturists should have a high skill level and up-to-date knowledge in handling forest stands based on forest ecology and silvics that must be adapted to economic factors, including financial, institutional and social concerns (Ford-Robertson, 1971; Helms, 1998; Nyland, 2016).

7.2. Silviculture: Existing regulations

The Indonesian Selective Cutting System (TPI) was the first Indonesian silviculture system for logging in concession areas in a natural mixed forest. The procedure required cutting only the harvestable trees, i.e. the commercial timber species, and confirming a diameter limit. However, the logging companies only complied with the minimum cutting diameter while paying little attention to reinvigorating the residual stand, maintaining future trees or enriching regeneration as required by the system (Yeom & Chandrasekharan, 2001).

In 1989, the Ministry of Forestry updated the TPI to the Indonesian Selective Cutting and Replanting System (TPTI), emphasising natural regeneration and enrichment planting. The TPTI requires a legal and definitive concession area, called a Forest Management Unit (FMU), to ensure sustainability. An FMU can cover one or more stands as long as they do not vary much in timber volume. Based on a 35-year cutting cycle, an FMU applies the TPTI depending on Annual Allowable Cutting (AAC), determined by the Forestry Ministry after assigning the targeted (productive) area and evaluating the estimated harvestable volume.

Under the TPTI, the prescribed minimum cutting limit varies depending on the type of forest function, i.e. 50cm DBH for production forest, 60cm DBH for limited production forest, and 40cm DBH for production forest in swamp areas. In each forest type, the TPTI requires retaining at least 25 commercially valuable trees per hectare. The diameter of these residual trees should be in the range of 20–50 cm DBH in production forest, 20–60 cm DBH in limited production forest and

20–40 cm DBH in the swamp forest. The concessionary should regularly report the natural regeneration, enrichment planting and stand maintenance (e.g. pruning future trees).

Furthermore, the Ministry of Forestry introduced new techniques such as Reduced Impact Logging (RIL) and Intensive Silviculture (SILvikultur INtensif/ SILIN) in TPTI. Their goals are to reduce forest damage, increase stand productivity and improve planting intensity in logged areas. The approaches are not mandatory, but their influence on forest product certification is significant since the aims are to reach sustainability in economic, social and production concerns.

However, some handicaps to ensuring sustainable yields have remained during the implementation of TPTI. In most cases, TPTI has had a weak impact on controlling harvesting intensity, regardless of RIL or the conventional logging method applied in the system, which often damaged trees of the remaining stand and thus led to higher tree mortality (Sist *et al.*, 1998; Sist & Nguyen-Thé, 2002).

7.3. Implications of species performance for silviculture approach

Considering the threat to natural forests caused by conventional logging, the authorities in the tropics have designated forestry legislation that might transform the paradigm of timber yield orientation to sustainability. For example, the philosophy of sustainable forest management (SFM) in the Indonesian production forest estate is to ensure that its management units will comply with sustainability requirements in ecological, social and production aspects included in its strategic planning for the long-term period.

The Ministry of Forestry imposed the criteria and indicators for SFM through Regulation No. 4795/Kpts-II/2002, issued on 3 June 2002. To control timber harvesting in the production forest, the Ministry has issued Regulation No. P. 11/Menhut-II/2009 to strengthen the TPTI; when this is implemented in combination with RIL and SILIN, today's logging practices will change from predatory to sustainable.

Studies of authority regulations on SFM in Southeast Asia and Amazonia are in complete agreement and admit that the commercial value of most timber species depletes within three harvesting cycles (Kammesheidt *et al.*, 2001; Hall *et al.*, 2003; Sist *et al.*, 2003a; Dauber *et al.*, 2005; van Gardingen *et al.*, 2006; de Freitas & Pinard, 2008; Forshed *et al.*, 2008; Grogan *et al.*, 2008; Peña-Claros *et al.*, 2008; Schulze *et al.*, 2008; Kukkonen & Hohnwald, 2009; Gaveau *et al.*, 2013). Moreover, the conclusion is that the protocols of SFM for tropical forests have some limitations, i.e. short rotation cycles of typically 25–35 years, a minimum felling diameter (mostly 50 cm DBH) too small to preserve an adequate density of adults and control harvest intensity (around eight trees per ha), and inadequate seed tree retention. In addition, these studies have confirmed that because of significant trait differences in commercial timbers, it is challenging to control timber species depletion and residual-value timber retention by applying a single minimum cutting diameter. Therefore, adaptable protocols that mimic species-specific characteristics, including local population structure, growth rates and relative densities, are urgently needed (Sist *et al.*, 2003b; Dauber *et al.*, 2005; de Freitas & Pinard, 2008; Schulze *et al.*, 2008; Herault *et al.*, 2010).

Maximising access to light and nutrients for high-value timber trees, seedlings, and saplings is the sustainable sense of the art and science of tropical silviculture. The presented study demonstrates that in the steep stand, where light is more available through gaps in the contours or single tree gaps, the basal areas and diversity were greater than in the flat stands. This finding is consistent with the statement that primary forest tree species respond swiftly to competitive release (tree or branch fall) and unexpected light provided in canopy gaps by augmenting their diameter increment (Herault *et al.*, 2010). Accordingly, removing competing trees and vines is essential. Regardless, treatments must usually be reapplied at intervals to maintain a growth advantage (Peña-Claros *et al.*, 2008; Herault *et al.*, 2010). However, the silviculture models developed for tropical forests have failed to meet the goal of sustainable timber production at an execution level. Even though post-logging silviculture treatments have been widely implemented, including in Indonesia and Malaysia, they have failed to achieve the objective of sustainable timber production or forest conservation (Kuusipalo *et al.*, 1997; Hall *et al.*, 2003; Sist *et al.*, 2003a).

This study reveals some indications that TPTI will not provide sustainable yields after the second cutting cycle due to outcompeting of slow-growth shade-tolerant species and non-pioneer light-demanding species (i.e. high-value timbers) during the recovery of the logged stand. This result is in accordance with many studies that have demonstrated that high-value timber species are unlikely to regenerate adequately on their own after SFM logging. Pioneer vegetation prefers large gaps left by timber extraction and tends to colonise the site. Although the saplings of many tree species are capable of surviving long periods of suppression until the canopy opens again through the senescence of pioneers, suppressed stems do not grow at rates that will result in commercial-sized trees in the third harvest cycle. Alterations in regeneration under the current logging protocols will provoke tropical forests to regenerate naturally with a large composition of light-wooded tree species (non-commercial or low commercial value species), whereas dense-wood, high-value timber species will experience severe population declines (Kuusipalo *et al.*, 1997; Hall *et al.*, 2003; Park *et al.*, 2005; Valle *et al.*, 2007; Schulze, 2008).

Even though recent studies show the success of enrichment planting of pre-grown seedlings and their subsequent tending in small logging gaps as potential recovery of commercial species after sustainable-managed harvesting (Kuusipalo *et al.*, 1997; Park *et al.*, 2005; Grogan *et al.*, 2008; Hall, 2008; Schulze, 2008), not many improvements have been included in currently components of RIL or SFM, or as criteria for forest certification under FSC; e.g. supplementary silvicultural treatments or other selection of harvesting protocol required in augmenting the growing increment and abundance performances of non-pioneer shade-tolerant and -intolerant tree species after selective logging (Zimmerman and Kormos, 2012). In general, the enrichment-planting guidance is not needed for dipterocarp stand in Southeast Asia because of its abundance in regeneration, but the requirement for monitoring and controlling established seedlings in small gaps is the same as in Africa and Amazonia (Kuusipalo *et al.*, 1997; Sist *et al.*, 2003b). Moreover, the re-growing potential of timber trees in single-tree-size logging gaps that mimic the natural conditions for primary forest tree regeneration is vital in sustainable silviculture practice (Kuusipalo *et al.*, 1997; Sist *et al.*, 2003b; Schulze, 2008; Herault *et al.*, 2010). In addition, periodically thinning of competitive and suppressive pioneers at least the first few years can allow seedlings to successfully establish and develop in logging gaps (Peña-Claros et al., 2008). At last, reducing the harvesting intensity through the implementation of species-specific minimum cutting size (Cam, 2015) and extending the harvesting (forest management unit) area may potentially reach the production of commercial-sized timber in logging gaps by the third harvest, that is, the silvicultural goal now recommended by most tropical forestry researchers (Park *et al.*, 2005; Peña-Claros *et al.*, 2008; Schulze, 2008).

In conclusion, to prevent the continuation of timber productivity and ecological integrity in the Papuan logging concessionary, the species-based treatments that mimic its natural traits in an intact forest community are highly pivotal in silviculture protocol. The enrichment planting of pre-grown seedlings and their subsequent tending in small logging gaps is a must for outcompeting high-value timbers of slow-growth shade-tolerant and non-pioneer light-demanding species. Moreover, reducing the harvesting intensity below eight trees per ha with an indirect approach of species-based cutting limit should be implemented for all timber species (e.g. greater diameter cutting limit for giant canopy species with absent or fewer small-size individuals). Additionally, enlarging the targeted harvesting area to maintain production levels that potentially drop due to lowering the cutting intensity should be adopted in the application procedure of concession permit. Finally, these proposed silvicultural techniques cannot neglect economic, institutional and social concerns.

8. Summary

Forest degradation is currently a vital issue in Indonesian tropical forests, and the logging-driven loss of forest cover has emerged as a major concern in Papuan production forests. Most logging activities have failed to achieve sustainability in production and ecological integrity, even though the timber companies have applied the sustainability protocols of the national silviculture system. The national protocol has been recognised as too general for local adaptations, such as in Indonesia, which has many distinct types of production forests. Therefore, this study investigated the specific characteristics of Papuan's intact forest (including species composition, diversity, structure, regeneration and species light requirements) and to what extent logging impacted the stand characteristics in logged forests. Based on the differences in characteristics between intact and logged forests, the study evaluated tree species performances, highlighted key problems and recommended a better silviculture approach for the Indonesian silviculture system.

Tree species composition, diversity and structural characteristics of intact forests

The two Papuan logging concessions' primary forest characteristics varied greatly between locations (Tunas in the southern and Bonggo in the northern lowlands) and between stands with different slopes (the flat and steep stands).

There were 143 species in 43 families identified in this study; 90 species in 38 families and 72 species in 31 families were recorded at the Tunas and Bonggo sites, respectively. The top-ranked species and families in the Tunas forest were *Vatica rassak, Anisoptera polyandra, Hopea papuana* (Dipterocarpaceae), *Litsea timoriana, Beilsmiedia* sp. (Lauraceae), *Blumeodendron amboinicum, Pimelo-dendron amboinicum, Medusanthera polot* (Euphorbiaceae), *Canarium asperum, Dacryodes* sp. (Burseraceae), *Syzygium versteegii* and *Syzygium anomalum* (Myrtaceae). *Intsia bijuga* (Fabaceae), *Anisoptera polyandra* (Dipterocarpaceae), *Pometia pinata* (Sapindaceae), *Homalium foetidum* (Salicaceae), *Celtis riges-cens* (Canabaceae), *Myristica tubiflora, Myristica sulcata* (Myristicaceae), *Syzy-*

gium anomalum (Myrtaceae), *Litsea ledermannii* (Lauraceae), *Pimelodendron amboinicum* (Euphorbiaceae), and *Canarium asperum* (Burseraceae) were the important species and families in the Bonggo forest.

Even though the two forests had similar high tree diversities, they performed differently for various measures of diversity: The Tunas forest was typically higher in tree richness, whilst the Bonggo forest was greater in heterogeneity. Tree richness ranged from 87 species in the flat area to 89 species at the steep site in the Tunas forest, whereas the richness in Bonggo ranged between 63 and 67 species at the flat and steep sites, respectively. In addition, the species rank abundance indicated that the Tunas forest was higher in tree rareness (log α = 23 species for flat and steep stands) than the Bonggo forest, which had lower tree rareness (log α =14 and 16 species, respectively, for flat and steep stands). In the case of heterogeneity, the Bonggo stands were more equivalent in common species; 54% – 62% of species had relatively similar values for the abundance of individuals (Shannon exp. $e^{H'}$ = 34 – 41 species) compared to the Tunas stands, which had 55% – 56% species with equal abundances of individuals ($e^{H'}$ = 48 – 50 species). The individual-based inverse Simpson's index showed more varied proportions of common species in the Bonggo stands (1/D = 23% and 31%, respectively) than in Tunas (1/D = 27% and 30%, respectively). Moreover, when the heterogeneity indices were based on species' basal area (BA) instead of individual-based abundance, the $e^{H'}$ and 1/D of all Tunas stands and the Bonggo flat stand increased significantly, while those of the Bonggo steep stand indicated relatively stable $e^{H'}$ and decreasing 1/D. The findings on tree richness and heterogeneity implied that the Tunas and Bonggo forests might have different historical distributions of species and development.

The findings for structural characteristics indicated that the Bonggo had stands typically higher in tree density than the Tunas. In the Bonggo stands, the number of trees ranged from 402 stems ha^{-1} in the flat stand to 493 stems ha^{-1} in the steep stand with respective basal areas between 27 and 35 m^{2}ha^{-1}. In the Tunas stands, the number of trees varied between 269 and 319 stems ha^{-1} with respective basal areas of 22 and 25 m^{2}ha^{-1}. The distribution of tree diameter sizes indicated that the Bonggo forests were composed of a high percentage (47.51% in the steep and 43.59% in the flat areas) of trees in the smallest class (DBHs<20 cm) and few trees in long-tailed large diameter size classes (DBH 50 cm - 125 cm). Similarly, in the Tunas forest, a high proportion of trees (36.50% in the steep stand and 36.22% in the flat stand) were concentrated in the smallest diameter classes, and a very small proportion occurred in the short-tailed large-diameter classes (DBH 50 cm – 90 cm).

Further results showed that the mean heights of the Bonggo stands were 17.34 m (steep area) and 16.17 m (flat area). These were higher than the Tunas stands (15.62 m and 15.08 m). Further analysis of height variables (mean height of the 100 tallest trees, 20% of tallest trees and Lorey's mean height) implied that the giant trees and high-density canopy population most likely prevented more light penetration into the forest canopy at Bonggo than at Tunas. The restriction of light penetration resulted in high populations of stems in the lower canopy layer of Bonggo (50.87% and 63.66% of the total stems) and in the middle layer of Tunas (62.73% and 65.36%). The Chapman–Richards growth function showed more convincingly that competition for light resources among trees in the lower canopy was less intense in the Tunas stands than in the Bonggo stands, based on their transition points at 17 m and 28 m in height.

Regeneration in intact forest

Not all canopy species exhibited natural regeneration in the intact forest. The Chao–Jaccard similarity index showed that no more than 65% of canopy species were present as seedlings and 72% as saplings in the Bonggo forest. The canopy species' regeneration percentages were lower in the Tunas forest; 60% and 67% of the canopy species regenerated as seedlings and saplings, respectively. Competition for sunlight became the major factor in preventing regeneration in both forests.

There were 50 seedlings and 59 sapling species in the Tunas forest and 35 seedlings and 47 sapling species in the Bonggo forest. *Vatica rassak, Litsea timoriana, Garcinia celebica,* and *Blumeodendron amboinicum* were the top-ranked species in the Tunas forest. *Homalium foetidum, Celtis rigescens, Anisoptera polyandra, Canarium asperum, and Calophyllum inophyllum* were the dominant species in the Bonggo forest. All these high-ranked species at the regeneration level were also the most important species at the tree level. Those absent from the regeneration level mostly belonged to less important or rare species, except for *Intsia bijuga* in the Bonggo forest.

The number of seedlings in the Tunas forest was found to be 28,250 ha^{-1}, with the number of stems decreasing to 6,080 ha^{-1} for the sapling category. In the Bonggo forest, the number of seedlings was 23,750 ha^{-1}, and the number of stems decreased to 3,720 ha^{-1} in the sapling category. Saplings distribution in its diameter classes (1.5 cm<DBH<10 cm) indicated high recruitment of juveniles to tree level (DBH>10 cm), representing fast-growing regeneration in the Tunas forest. By contrast, a stagnant supply of regeneration stocks was found in the Bonggo forest, representing a slow-growing understorey. Differences in sunlight penetration to the canopy layers of both forests were most likely the main reason for these observations.

<u>Logging impact on tree species composition, diversity and stand structure</u>

More than half of the 117 species in the Tunas forest were commercial timbers; 23 (19.7%) were high-value species, and 41 (35%) were mixed commercial species. In the Bonggo forest, 29 (34.5%) out of 84 species were high-value commercials, and the same percentage (34.5%) were mixed commercials.

Most of the dominant species in both forests were commercial timbers. For example, *Vatica rassak, Anisoptera polyandra, Canarium asperum, Hopea papuana* (high-valued group), *Litsea timoriana* and *Syzygium versteegii* (mixed group) were the dominant species in the Tunas forest, and *Intsia bijuga, Homalium foetidum, Celtis rigescens, Anisoptera polyandra* (high-value timbers), *Litsea ledermannii, Myristica sulcata, Myristica tubiflora,* and *Syzygium anomalum* (mixed commercials) were the most important species in the Bonggo forest. Other commercial species were categorised as less important and rare species in both forests.

Logging will potentially change the species composition of both forests in the future. The findings of this study suggested that the appearance of significant pioneer species in Tunas forest (but not in Bonggo forest) after the first cutting cycle is one reason for this, and the extinction of *Anisoptera polyandra, Hopea papuana* (in Tunas), *Intsia bijuga* (in Bonggo) and other less-important commercial species after third-cycle harvesting is another reason.

A non-significant logging-impacted reduction in tree species diversity was reported in the Tunas and Bonggo forests at a short time interval (8 years) following timber harvesting. However, the study predicted that if today's destruction and ecological changes related to logging occurred again in the next cutting cycle, this would negatively alter the species' relative abundance, thus downgrading the future degree of tree diversity.

The current logging practices left only 9% of stems in Tunas and 41% of stems in Bonggo from the initial harvestable timber stocks; the removed timbers included more than half of the initial commercial species lost in both forests. The harvesting activity also killed the future crops and left 85% of stems and 66% of species in Tunas, and 38% of stems and 55% of species in Bonggo. These results are consistent with the expectation that high harvesting intensity would cause a massive change in logged stand structure if the current logging protocols, such as a fixed minimum diameter cutting (50 cm) used for all species, were applied.

<u>Grouping species based on light requirements</u>

The study found five species groups based on the slope of their diameter distribution (b-value), which varied from +2.25 to −67.80 with a presumption that more negative values indicated shade-tolerant species and high positive values indicated light-demanding species. Another finding also showed a robust exponential correlation between the b-value and the stable N isotope ratio (δ^{15}N) and total N in the xylem tissue of these species. This is consistent with the study's expectation that nitrate reductase activity plays an important role in determining the values of δ^{15}N and total N in the xylem tissue. These values varied markedly between nutrient-conservative species, which can grow abundantly with limited light illumination, and non-nutrient–conservative species, which mainly rely on light for photosynthesis activity.

Based on the correlation between a species' b-value and its δ^{15}N and total N in the xylem tissue, the study successfully characterised 50 and 53 species from Tunas and Bonggo forests, respectively, into their light requirement groups: light-demanding, opportunist and shade-tolerant.

<u>The implications of species performance for silvicultural approaches</u>

The findings for tree species characteristics in intact and logged forests agree with other studies that it is challenging to control timber species depletion and residual timber value retention by applying a single minimum cutting diameter because of significant trait differences in commercial timbers. Therefore, adaptable protocols that consider species-specific characteristics, including local population structure, growth rates, relative densities and species light requirements, are urgently needed. A combination of species-specific minimum cutting diameters and enlargement of forest management units should be included in protocols to reduce and control the harvesting intensity.

Zusammenfassung

Walddegradation ist derzeit ein wichtiges Thema in den indonesischen Tropenwäldern, und der durch den Holzeinschlag verursachte Verlust der Waldbedeckung hat sich in den Produktionswäldern Papuas zu einem großen Problem entwickelt. Die meisten Abholzungsaktivitäten haben es nicht geschafft, Nachhaltigkeit in der Produktion und ökologische Integrität zu erreichen, obwohl die Holzfirmen die Nachhaltigkeitsprotokolle des nationalen Waldbausystems angewendet haben. Das nationale Protokoll hat sich als zu allgemein für lokale Anpassungen erwiesen, wie z.B. in Indonesien, das viele verschiedene Arten von Produktionswäldern hat. Daher untersuchte diese Studie die spezifischen Charakteristika intakter Wälder in Papua (einschließlich Artenzusammensetzung, Diversität, Struktur, Regeneration und Lichtbedürfnisse der Arten) und inwieweit die Abholzung die Bestandseigenschaften in abgeholzten Wäldern beeinflusste. Basierend auf den Unterschieden in den Charakteristika zwischen intakten und abgeholzten Wäldern, bewertete die Studie die Leistungen der Baumarten, hob Schlüsselprobleme hervor und empfahl einen besseren waldbaulichen Ansatz für das indonesische Waldbausystem.

<u>Baumartenzusammensetzung, Diversität und Strukturmerkmale intakter Wälder</u>

Die Primärwaldcharakteristika der beiden papuanischen Forstkonzessionen variierten stark zwischen den Standorten (Tunas im südlichen und Bonggo im nördlichen Tiefland) und zwischen Beständen mit unterschiedlichen Neigungen (den flachen und steilen Beständen).

In dieser Studie wurden 143 Arten in 43 Familien identifiziert; 90 Arten in 38 Familien und 72 Arten in 31 Familien wurden an den Standorten Tunas bzw. Bonggo erfasst. Die ranghöchsten Arten und Familien im Tunas-Wald waren *Vatica rassak, Anisoptera polyandra, Hopea papuana* (Dipterocarpaceae), *Litsea timoriana, Beilsmiedia* sp. (Lauraceae), *Blumeodendron amboinicum, Pimelodendron amboinicum, Medusanthera polot* (Euphorbiaceae), *Canarium asperum, Dacryodes* sp. (Burseraceae), *Syzygium versteegii* und *Syzygium anomalum*

(Myrtaceae). *Intsia bijuga* (Fabaceae), *Anisoptera polyandra* (Dipterocarpaceae), *Pometia pinata* (Sapindaceae), *Homalium foetidum* (Salicaceae), *Celtis rigescens* (Canabaceae), *Myristica tubiflora*, *Myristica sulcata* (Myristicaceae), *Syzygium anomalum* (Myrtaceae), *Litsea ledermannii* (Lauraceae), *Pimelodendron amboinicum* (Euphorbiaceae), und *Canarium asperum* (Burseraceae) waren die wichtigen Arten und Familien im Bonggo-Wald.

Obwohl die beiden Wälder eine ähnlich hohe Baumdiversität aufwiesen, schnitten sie bei verschiedenen Maßstäben für die Diversität unterschiedlich ab: Der Tunas-Wald wies typischerweise einen höheren Baumreichtum auf, während der Bonggo-Wald eine größere Heterogenität aufwies. Der Baumreichtum reichte im Tunas-Wald von 87 Arten im flachen Bereich bis zu 89 Arten am steilen Standort, während der Reichtum in Bonggo zwischen 63 und 67 Arten an den flachen bzw. steilen Standorten lag. Außerdem zeigte die Artenranghäufigkeit, dass der Tunas-Wald eine höhere Baumhäufigkeit aufwies (log α = 23 Arten für flache und steile Bestände) als der Bonggo-Wald, der eine geringere Baumhäufigkeit aufwies (log α =14 bzw. 16 Arten für flache und steile Bestände). Bei der Heterogenität waren die Bonggo-Bestände gleichwertiger in Bezug auf häufige Arten; 54% - 62% der Arten hatten relativ ähnliche Werte für die Abundanz von Individuen (Shannon exp. $e^{H'}$ = 34 - 41 Arten) im Vergleich zu den Tunas-Beständen, die 55% - 56% Arten mit gleichen Abundanzen von Individuen hatten ($e^{H'}$ = 48 - 50 Arten). Der individuen-basierte inverse Simpson's Index zeigte in den Bonggo-Beständen unter-schiedlichere Anteile an häufigen Arten (1/D = 23% bzw. 31%) als in Tunas (1/D = 27% bzw. 30%). Wenn die Heterogenitätsindizes auf der Basalfläche (BA) der Arten anstelle von individuenbasierter Abundanz basierten, stiegen $e^{H'}$ und 1/D aller Tunas-Bestände und des flachen Bonggo-Bestands signifikant an, während die des steilen Bonggo-Bestands relativ stabile $e^{H'}$ und abnehmende 1/D zeigten. Die Befunde zum Baumreichtum und zur Heterogenität deuteten darauf hin, dass die Tunas- und Bonggo-Wälder möglicherweise unterschiedliche historische Verteilungen von Arten und Entwicklungen aufweisen.

Die Ergebnisse für die strukturellen Merkmale zeigten, dass die Bonggo-Bestände typischerweise eine höhere Baumdichte aufwiesen als die Tunas. In den Bonggo-Beständen reichte die Anzahl der Bäume von 402 Stämmen ha^{-1} im flachen Bestand bis zu 493 Stämmen ha^{-1} im steilen Bestand mit entsprechenden Grundflächen zwischen 27 und 35 m^{2}ha^{-1}. In den Tunas-Beständen variierte die Anzahl der Bäume zwischen 269 und 319 Stämmen ha^{-1} mit entsprechenden Grundflächen von 22 und 25 m^{2}ha^{-1}. Die Verteilung der Baumdurchmessergrößen zeigte, dass die Bonggo-Wälder aus einem hohen Prozentsatz (47,51% in den steilen und 43,59% in den flachen Gebieten) von Bäumen der kleinsten Klasse (DBH<20 cm) und wenigen Bäumen der großen Durchmessergrößenklassen (DBH 50 cm - 125 cm) bestanden. In ähnlicher Weise konzentrierte sich im Tunas-Wald ein hoher Anteil von Bäumen (36,50% im steilen Bestand und 36,22% im flachen Bestand) in den kleinsten Durchmesserklassen und ein sehr geringer Anteil in den kurzschwänzigen großen Durchmesserklassen (DBH 50cm - 90 cm).

Weitere Ergebnisse zeigten, dass die mittleren Höhen der Bonggo-Bestände 17,34 m (steiler Bereich) und 16,17 m (flacher Bereich) betrugen. Diese waren höher als bei den Tunas-Beständen (15,62 m und 15,08 m). Eine weitere Analyse der Höhenvariablen (mittlere Höhe der 100 höchsten Bäume, 20% der höchsten Bäume und die mittlere Höhe von Lorey) ergab, dass die Baumriesen und die hohe Dichte des Baumkronenbestandes höchstwahrscheinlich mehr Lichteinfall in die Waldkronen in Bonggo als in Tunas verhinderten. Dies führte zu einem hohen Bestand an Stämmen in der unteren Kronenschicht von Bonggo (50,87% und 63,66% der gesamten Stämme) und in der mittleren Schicht von Tunas (62,73% und 65,36%). Die Chapman-Richards-Wachstumsfunktion zeigte überzeugender, dass die Konkurrenz um Lichtressourcen zwischen den Bäumen im unteren Kronendach in den Tunas-Beständen weniger intensiv war als in den Bonggo-Beständen, basierend auf deren Übergangspunkten bei 17 m und 28 m Höhe.

<u>Regeneration in intaktem Wald</u>

Nicht alle Baumkronenarten wiesen im intakten Wald eine natürliche Verjüngung auf. Der Chao-Jaccard-Ähnlichkeitsindex zeigte, dass nicht mehr als 65% der Baumkronenarten als Sämlinge und 72% als Setzlinge im Bonggo-Wald vorhanden waren. Der Prozentsatz der Regeneration der Baumkronenarten war im Tunas-Wald niedriger; 60% und 67% der Baumkronenarten regenerierten sich als Sämlinge bzw. Schösslinge. Die Konkurrenz um Sonnenlicht war der Hauptfaktor, der die Regeneration in beiden Wäldern verhinderte.

Es gab 50 Sämlings- und 59 Schösslingsarten im Tunas-Wald und 35 Sämlings- und 47 Schösslingsarten im Bonggo-Wald. *Vatica rassak, Litsea timoriana, Garci-nia celebica* und *Blumeodendron amboinicum* waren die am häufigsten vorkom-menden Arten im Tunas-Wald. *Homalium foetidum, Celtis rigescens, Anisoptera polyandra, Canarium asperum und calophyllum inophyllum* waren die dominierenden Arten im Bonggo-Wald. Alle diese hochrangigen Arten auf der Verjüngungsebene waren auch die wichtigsten Arten auf der Baumebene. Diejenigen, die auf der Verjüngungsebene fehlten, gehörten meist zu weniger wichtigen oder seltenen Arten, mit Ausnahme von *Intsia bijuga* im Bonggo-Wald.

Die Anzahl der Setzlinge im Tunas-Wald wurde mit 28.250 ha^{-1} ermittelt, wobei die Anzahl der Stämme auf 6.080 ha^{-1} in der Kategorie Setzlinge zurückging. Im Bonggo-Wald betrug die Anzahl der Setzlinge 23.750 ha^{-1}, und die Anzahl der Stämme sank auf 3.720 ha^{-1} in der Setzlingskategorie. Die Verteilung der Setzlinge in ihren Durchmesserklassen (1,5 cm<DBH< 10 cm) deutete im Tunas-Wald auf eine hohe Rekrutierung von Jungbäumen auf Baumebene (DBH>10cm) hin, die eine schnell wachsende Verjüngung darstellen. Im Gegensatz dazu wurde im Bonggo-Wald ein stagnierender Nachschub an Verjüngungsbeständen gefunden, was langsam wachsendes Unterholz repräsentiert. Unterschiede in der Sonnen-einstrahlung in die Kronenschichten der beiden Wälder waren wahrscheinlich der Hauptgrund für diese Beobachtungen.

<u>Auswirkungen des Holzeinschlags auf Baumartenzusammensetzung, -vielfalt und Bestandsstruktur</u>

Mehr als die Hälfte der 117 Arten im Tunas-Wald waren Handelshölzer; 23 (19,7%) waren hochwertige Arten und 41 (35%) waren gemischte Handelsarten. Im Bonggo-Wald waren 29 (34,5%) von 84 Arten hochwertige Nutzhölzer, und der gleiche Prozentsatz (34,5%) waren gemischte Nutzhölzer.

Die meisten der dominierenden Arten in beiden Wäldern waren Nutzhölzer. Zum Beispiel waren *Vatica rassak, Anisoptera polyandra, Canarium asperum, Hopea papuana* (hochwertige Gruppe), *Litsea timoriana* und *Syzygium versteegii* (gemischte Gruppe) die dominierenden Arten im Tunas-Wald, und *Intsia bijuga, Homalium foetidum, Celtis rigescens, Anisoptera polyandra* (hochwertige Hölzer), *Litsea ledermannii, Myristica sulcata, Myristica tubiflora* und *Syzygium anomalum (gemischte* Handelsarten) die wichtigsten Arten im Bonggo-Wald waren. Andere Nutzholzarten wurden in beiden Wäldern als weniger wichtige und seltene Arten eingestuft.

Die Abholzung wird möglicherweise die Artenzusammensetzung beider Wälder in der Zukunft verändern. Die Ergebnisse dieser Studie deuten darauf hin, dass das Auftreten bedeutender Pionierarten im Tunas-Wald (aber nicht im Bonggo-Wald) nach dem ersten Hiebszyklus ein Grund dafür ist, und das Aussterben von *Anisoptera polyandra, Hopea papuana* (in Tunas), *Intsia bijuga* (in Bonggo) und anderen weniger wichtigen kommerziellen Arten nach dem dritten Erntezyklus ein weiterer Grund ist.

In den Wäldern Tunas und Bonggo wurde in einem kurzen Zeitintervall (8 Jahre) nach der Holzernte eine nicht signifikante, durch den Holzeinschlag verursachte Verringerung der Baumartenvielfalt festgestellt. Die Studie prognostizierte jedoch, dass, wenn die heutigen Zerstörungen und ökologischen Veränderungen im Zusammenhang mit der Holzernte im nächsten Einschlagszyklus erneut auftreten, dies die relative Häufigkeit der Arten negativ verändern und damit den zukünftigen Grad der Baumvielfalt herabsetzen würde.

Die gegenwärtigen Abholzungspraktiken ließen nur 9% der Stämme in Tunas und 41% der Stämme in Bonggo von den ursprünglich erntbaren Holzvorräten übrig; die entfernten Hölzer beinhalteten mehr als die Hälfte der ursprünglich kommerziell genutzten Arten, die in beiden Wäldern verloren gingen. Die Erntetätigkeit tötete auch die zukünftigen Bestände und hinterließ 85% der Stämme und 66% der Arten in Tunas und 38% der Stämme und 55% der Arten in Bonggo. Diese Ergebnisse stimmen mit der Erwartung überein, dass eine hohe Ernteintensität zu einer massiven Veränderung der Bestandesstruktur führen würde, wenn die aktuellen Abholzungsprotokolle, wie z. B. ein fester Mindestdurchmesserschnitt (50 cm) für alle Arten, angewendet würden.

<u>Gruppierung von Arten auf der Grundlage von Lichtanforderungen</u>

Die Studie fand fünf Artengruppen basierend auf der Steigung ihrer Durchmesser-verteilung (b-Wert), die von +2,25 bis -67,80 variierte, mit der Vermutung, dass negativere Werte auf schattentolerante Arten und hohe positive Werte auf lichtbedürftige Arten hinweisen. Ein weiteres Ergebnis zeigte auch eine robuste exponentielle Korrelation zwischen dem b-Wert und dem stabilen N-Isotopenverhältnis ($\delta^{15}N$) und dem Gesamt-N im Xylemgewebe dieser Arten. Dies steht im Einklang mit der Erwartung der Studie, dass die Nitratreduktase-Aktivität eine wichtige Rolle bei der Bestimmung der Werte von $\delta^{15}N$ und Gesamt-N im Xylemgewebe spielt. Diese Werte variierten deutlich zwischen nährstoffkonservativen Arten, die bei begrenzter Lichtbeleuchtung üppig wachsen können, und nicht-nährstoffkonservativen Arten, die für die Photosynthese-aktivität hauptsächlich auf Licht angewiesen sind.

Basierend auf der Korrelation zwischen dem b-Wert einer Art und ihrem $\delta^{15}N$ und Gesamt-N im Xylemgewebe wurden in der Studie 50 bzw. 53 Arten aus Tunas- und Bonggo-Wäldern erfolgreich in ihre Lichtbedarfsgruppen eingeordnet: lichtbedürftig, opportunistisch und schattentolerant.

Die Ergebnisse für Baumarteneigenschaften in intakten und abgeholzten Wäldern stimmen mit anderen Studien überein, dass es eine Herausforderung ist, die Holzartenverarmung und den Erhalt des Restholzwertes durch die Anwendung eines einzigen Mindestschnittdurchmessers zu kontrollieren, da es signifikante Merkmalsunterschiede bei kommerziellen Hölzern gibt. Daher werden dringend anpassungsfähige Protokolle benötigt, die artenspezifische Merkmale wie die lokale Populationsstruktur, Wachstumsraten, relative Dichten und Lichtbedürfnisse der Arten berücksichtigen. Eine Kombination aus artspezifischen Mindesthiebs-durchmessern und einer Vergrößerung der Waldbewirtschaftungseinheiten sollte in die Protokolle aufgenommen werden, um die Ernteintensität zu reduzieren und zu kontrollieren.

References

Abood SA, Lee JSH, Burivalova Z, Garcia-Ulloa J, Koh LP, 2015. Relative contributions of the logging, fiber, oil palm, and mining industries to forest loss in Indonesia. Conservation Letters 8 (1): 58–67.

Alder D, 1995. Growth modelling for mixed tropical forests. Oxford Forestry Institute, University of Oxford.

Allison A, 2007. Introduction to the Fauna of Papua. In A.J. Marshall and B.M. Beehler (Ed.). The Ecology of Papua Part I, The Ecology of Indonesia series, Vol. IV. (1): 479–494.

Anitha K, Joseph S, Chandran RJ, Ramasamy EV, Prasad SN, 2010. Tree species diversity and community composition in a human-dominated tropical forest of Western Ghats biodiversity hotspot, India. Ecological Complexity 7 (2): 217–224.

Ashton PS, Hall P, 1992. Comparisons of structure among mixed dipterocarp forests of north-western Borneo. Journal of Ecology 459–481.

Aslam M, Huffaker RC, 1982. In vivo nitrate reduction in roots and shoots of barley (Hordeum vulgare L.) seedlings in light and darkness. Plant Physiology 70 (4): 1009–1013.

Aslam M, Huffaker RC, Rains DW, Rao KP, 1979. Influence of light and ambient carbon dioxide concentration on nitrate assimilation by intact barley seedlings. Plant Physiology 63 (6): 1205–1209.

Asner GP, Broadbent EN, Oliveira PJC, Keller M, Knapp DE, Silva JNM, 2006. Condition and fate of logged forests in the Brazilian Amazon. Proceedings of the National Academy of Sciences 103 (34): 12947–12950.

Attridge TH, 1990. Light and plant responses: a study of plant photophysiology and the natural environment. Cambridge University Press.

Austin KG, Mosnier A, Pirker J, McCallum I, Fritz S, Kasibhatla PS, 2017. Shifting patterns of oil palm driven deforestation in Indonesia and implications for zero-deforestation commitments. Land Use Policy 69: 41–48.

Austin KG, Schwantes A, Gu Y, Kasibhatla PS, 2019. What causes deforestation in Indonesia? Environmental Research Letters 14 (2): 1–9.

Babweteera F, Brown N, 2010. Spatial patterns of tree recruitment in East African tropical forests that have lost their vertebrate seed dispersers. Journal of Tropical Ecology 26 (2): 193–203.

Beaudet M, Messier C, Paré D, Brisson J, Bergeron Y, 1999. Possible mechanisms of sugar maple regeneration failure and replacement by beech in the Boisé-des-Muir old-growth forest, Québec. Ecoscience 6 (2): 264–271.

Beehler BM, 2007. Papuan terrestrial biogeography, with special reference to birds. In A.J. Marshall and B.M. Beehler (Ed.). The ecology of Papua Part I, The ecology of Indonesia series, Vol. IV. 196–206.

Bergeron Y, Harvey B, Leduc A, Gauthier S, 1999. Forest management guidelines based on natural disturbance dynamics: Stand- and forest-level considerations. The Forestry Chronicle 75 (1): 49–54.

Blaser J, Sarre A, Poore D, Johnson S, 2011. Status of tropical forest management 2011. ITTO technical series 38 (June): 376–384.

Bloom AJ, 1988. Ammonium and Nitrate as Nitrogen-Sources for Plant-Growth. Isi Atlas of Science-Animal & Plant Sciences 1 (1): 55–59.

Bloor JMG, Grubb PJ, 2003. Growth and mortality in high and low light: trends among 15 shade-tolerant tropical rain forest tree species. Journal of Ecology 91 (1): 77–85.

BPKH Wilayah X, 2009. Statistik Kehutanan Provinsi Papua Tahun 2008. Balai Pemantapan Kawasan Hutan (BPKH) Wilayah X, Jayapura-Indonesia.

Bray DB, Merino-Pérez L, Negreros-Castillo P, Segura-Warnholtz G, Torres-Rojo JM, Vester HFM, 2003. Mexico's community-managed forests as a global model for sustainable landscapes. Conservation Biology 17 (3): 672–677.

Broun AF, 1912. Sylviculture in the Tropics. Macmillan and Company, New York.

Brun R, 1969. Strukturstudien im gemaessigten Regenwald Suedchiles als Grundlage fuer Zustandserhebungen und Forstbetriebsplanung. University of Freiburg i. Br., Germany.

Bryan J, Shearman P, Ash J, Kirkpatrick JB, 2010. Estimating rainforest biomass stocks and carbon loss from deforestation and degradation in Papua New Guinea 1972–2002: best estimates, uncertainties and research needs. Journal of Environmental Management 91 (4): 995–1001.

Cain SA, 1938. The species-area curve. American Midland Naturalist 573–581.

Cam NV, 2015. Long-term effects of logging on structures and dynamics of moist evergreen forests in Kon Ha Nung, Central Highlands of Vietnam. Cuvillier Verlag, Göttingen,

Cannon CH, Peart DR, Leighton M, 1998. Tree species diversity in commercially logged Bornean rainforest. Science 281 (5381): 1366–1368.

Casson A, 2000. The hesitant boom: Indonesia's oil palm sub-sector in an era of economic crisis and political change. Center for International Forestry Research, Bogor, Indonesia.

Chao A, Hwang W-H, Chen YC, Kuo CY, 2000. Estimating the number of shared species in two communities. Statistica Sinica 10 (1): 227–246.

Chape S, Harrison J, Spalding M, Lysenko I, 2005. Measuring the extent and effectiveness of protected areas as an indicator for meeting global biodiversity targets. Philosophical Transactions of the Royal Society B: Biological Sciences 360 (1454): 443–455.

Chatterjee S, Hadi AS, 2015. Regression analysis by example. John Wiley & Sons, Chichester, UK.

Clark CJ, Poulsen, JR, Malonga R, ELKAN, Jr, PW, 2009. Logging concessions can extend the conservation estate for Central African tropical forests. Conservation Biology 23 (5): 1281–1293.

Cole DW, Rapp M, 1981. Elemental cycling in forest ecosystems. Dynamic properties of forest ecosystems 23: 341–409.

Colwell RK, 2013. EstimateS: Statistical estimation of species richness and shared species from samples. Version 9. http://purl. oclc. org/estimates.

Colwell RK, Chao A, Gotelli NJ, Lin S-Y, Mao CX, Chazdon RL, Longino JT, 2012. Models and estimators linking individual-based and sample-based rarefaction, extrapolation and comparison of assemblages. Journal of plant ecology 5 (1): 3–21.

Condit R, Hubbell SP, Foster RB, 1996. Assessing the response of plant functional types to climatic change in tropical forests. Journal of Vegetation Science 7 (3): 405–416.

Curran LM, Trigg SN, McDonald AK, Astiani D, Hardiono YM, Siregar P, Caniago I, Kasischke E, 2004. Lowland forest loss in protected areas of Indonesian Borneo. Science 303 (5660): 1000–1003.

Curtis JT, Mcintosh RP, 1950. The interrelations of certain analytic and synthetic phytosociological characters. Ecology 31 (3): 434–455.

Dauber E, Fredericksen TS, Peña M, 2005. Sustainability of timber harvesting in Bolivian tropical forests. Forest Ecology and Management 214 (1-3): 294–304.

Dawson TE, Mambelli S, Plamboeck AH, Templer PH, Tu KP, 2002. Stable isotopes in plant ecology. Annual Review of Ecology and Systematics 33 (1): 507–559.

de Freitas JV de, Pinard MA, 2008. Applying ecological knowledge to decisions about seed tree retention in selective logging in tropical forests. Forest Ecology and Management 256 (7): 1434–1442.

DeFries RS, Rudel T, Uriarte M, Hansen M, 2010. Deforestation driven by urban population growth and agricultural trade in the twenty-first century. Nature Geoscience 3 (3): 178–181.

Denslow JS, 1987. Tropical rainforest gaps and tree species diversity. Annual Review of Ecology and Systematics 18 (1): 431–451.

Denslow JS, 1995. Disturbance and diversity in tropical rain forests: the density effect. Ecological Applications 5 (4): 962–968.

Domingues TF, Martinelli LA, Ehleringer JR, 2007. Ecophysiological traits of plant functional groups in forest and pasture ecosystems from eastern Amazonia, Brazil. Plant Ecology 193 (1): 101–112.

Donato DC, Kauffman JB, Murdiyarso D, Kurnianto S, Stidham M, Kanninen M, 2011. Mangroves among the most carbon-rich forests in the tropics. Nature Geoscience 4 (5): 293–297.

Edwards DP, Laurance WF, 2013. Biodiversity despite selective logging. Science 339 (6120): 646–647.

Ekadinata S, van Noordwijk M, Budidarsono S, Dewi S, 2013. Hot spots in Riau, haze in Singapore: the June 2013 event analyzed. ASB Policy Brief (33): 6.

Ellenberg H, Nettels T, 2001. NHy-emission density triggers diversity of "typical" forest vascular plants. CBD Technical Series No. 3: Assesment, Conservation and Sustainable Use of Biodiversity 28–29.

Evans RD, Bloom AJ, Sukrapanna SS, Ehleringer, JR, 1996. Nitrogen isotope composition of tomato (Lycopersicon esculentum Mill. cv. T-5) grown under ammonium or nitrate nutrition. Plant, Cell & Environment 19 (11): 1317–1323.

Evans R, 2001. Physiological mechanisms influencing plant nitrogen isotope composition. Trends in Plant Science 6 (3): 121–126.

FAO, 1995. Forest Resources Assessment 1990: Global Synthesis: FAO Forestry Paper 124. Food and Agriculture Organization of the United Nations, Rome.

FAO, 2014. Criteria and Indicators - for sustainable forest management. Food and Agriculture Organization of the United Nations, Rome, Italy.

FAO-Unesco, 1974. Soil map of the world. Scale 1:5.000.000. Food and Agriculture Organization of the United Nations (FAO) - United Nations Educational, Scientific and Cultural Organization (Unesco), Paris.

Fisher B, Edwards DP, Larsen TH, Ansell FA, Hsu WW, Roberts CS, Wilcove DS, 2011. Cost-effective conservation: Calculating biodiversity and logging trade-offs in Southeast Asia. Conservation Letters 4 (6): 443–450.

Foley JA, Ramankutty N, Brauman KA, Cassidy ES, Gerber JS, Johnston M, Mueller ND, O'Connell C, Ray DK, West PC, 2011. Solutions for a cultivated planet. Nature 478 (7369): 337–342.

Ford-Robertson F, 1971. Terminology of forest science technology practice and products. Society of American Foresters, Washington, D.C.

Forshed O, Karlsson A, Falck J, Cedergren J, 2008. Stand development after two modes of selective logging and pre-felling climber cutting in a dipterocarp rainforest in Sabah, Malaysia. Forest Ecology and Management 255 (3-4): 993–1001.

Fox JC, Yosi CK, Nimiago P, Oavika F, Pokana JN, Lavong K, Keenan RJ, 2010. Assessment of Aboveground Carbon in Primary & Selectively Harvested Tropical Forest in Papua NewGuinea. Biotropica 42 (4): 410–419.

Fredericksen TS, 1999. Regeneration status of important tropical forest tree species in Bolivia: assessment and recommendations. Forest Ecology and Management 124 (2-3): 263–273.

Gandhi Y, Mitlöhner R, 2014. Tree species composition, diversity, and structure in Tunas logging concession area of Papua, Indonesia. Tropentag 2014, Book of Abstracts: Biophysical and Socio-economic Frame Conditions for the Sustainable Management of Natural Resources: International Research on Food Security, Natural Resource Management and Rural Development, Hamburg. 300.

Gaveau DL, Salim A, 2013. Research: nearly a quarter of June fires in Indonesia occurred in industrial plantations. Center for International Forestry and Research, Bogor, Indonesia.

Gaveau DL, Wich S, Epting J, Juhn D, Kanninen M, Leader-Williams N, 2009. The future of forests and orangutans (Pongo abelii) in Sumatra: predicting impacts of oil palm plantations, road construction, and mechanisms for reducing carbon emissions from deforestation. Environmental Research Letters 4 (3): 34013.

Gaveau DLA, Kshatriya M, Sheil D, Sloan S, Molidena E, Wijaya A, Wich S, Ancrenaz M, Hansen M, Broich M, 2013. Reconciling forest conservation and logging in Indonesian Borneo. PloS One 8 (8): e69887.

Gentry AH, 1988. Changes in plant community diversity and floristic composition on environmental and geographical gradients. Annals of the Missouri Botanical Garden 1–34.

Gibbs HK, Ruesch AS, Achard F, Clayton MK, Holmgren P, Ramankutty N, Foley JA, 2010. Tropical forests were the primary sources of new agricultural land in the 1980s and 1990s. Proceedings of the National Academy of Sciences 107 (38): 16732–16737.

Gotelli NJ, Graves GR, 1996. Null models in ecology. Smithsonian Institution Press, Washington, DC.

Gotelli NJ, Colwell RK, 2001. Quantifying biodiversity: procedures and pitfalls in the measurement and comparison of species richness. Ecology Letters 4 (4): 379–391.

Govindarajan S, Dietze M, Agarwal PK, Clark JS, 2004. A Scalable Simulator for Forest Dynamics. In Proceedings of the Twentieth Annual Symposium on Computational Geometry, June 9-11, 2004, Brooklyn, New York, USA pp. 106–115.

Grogan J, Jennings SB, Landis RM, Schulze M, Baima AMV, Lopes, J do Carmo A, Norghauer JM, Oliveira LR, Pantoja F, Pinto D, 2008. What loggers leave behind: impacts on big-leaf mahogany (*Swietenia macrophylla*) commercial populations and potential for post-logging recovery in the Brazilian Amazon. Forest Ecology and Management 255 (2): 269–281.

Guehl JM, Domenach A-M, Bereau M, Barigah TS, Casabianca H, Ferhi A, Garbaye J, 1998. Functional diversity in an Amazonian rainforest of French Guyana: a dual isotope approach (δ 15 N and δ 13 C). Oecologia 116 (3): 316–330.

Hall JS, 2008. Seed and seedling survival of African mahogany (Entandrophragma spp.) in the Central African Republic: implications for forest management. Forest Ecology and Management 255 (2): 292–299.

Hall JS, Harris DJ, Medjibe V, Ashton PMS, 2003. The effects of selective logging on forest structure and tree species composition in a Central African forest: implications for management of conservation areas. Forest Ecology and Management 183 (1-3): 249–264.

Handley LL, Raven JA, 1992. The use of natural abundance of nitrogen isotopes in plant physiology and ecology. Plant, Cell & Environment 15 (9): 965–985.

Hao Z, Zhang J, Song B, Ye J, Li B, 2007. Vertical structure and spatial associations of dominant tree species in an old-growth temperate forest. Forest Ecology and Management 252 (1-3): 1–11.

Hawthorne WD, Marshall CA, Juam MA, Agyeman VK, 2011. The impact of logging damage on tropical rainforest, their recovery and regeneration. An Annotated Bibliography 47–121.

Heck Jr KL, van Belle G, Simberloff D, 1975. Explicit calculation of the rarefaction diversity measurement and the determination of sufficient sample size. Ecology 56 (6): 1459–1461.

Helms JA, 1998. The dictionary of forestry. Society of American Foresters. Bethesda, MD, USA.

Henry HA, Aarssen LW, 1999. The interpretation of stem diameter–height allometry in trees: biomechanical constraints, neighbour effects, or biased regressions? Ecology Letters 2 (2): 89–97.

Herault B, Ouallet J, Blanc L, Wagner F, Baraloto C, 2010. Growth responses of neotropical trees to logging gaps. Journal of Applied Ecology 47 (4): 821–831.

Hill MO, 1973. Diversity and evenness: a unifying notation and its consequences. Ecology 54 (2): 427–432.

Hoering T, 1955. Variations of nitrogen-15 abundance in naturally occurring substances. Science 122 (3182): 1233–1234.

Holbrook NM, Putz FE, 1989. Influence of neighbors on tree form: effects of lateral shade and prevention of sway on the allometry of Liquidambar styraciflua (sweet gum). American Journal of Botany 76 (12): 1740–1749.

Hope GS, Hartemink AE, 2007. Soils of Papua. In A.J. Marshall and B.M. Beehler (Ed.). The ecology of Papua Part I, The ecology of Indonesia series, Vol. IV. 165–176.

Houghton RA, Hackler JL, 1999. Emissions of carbon from forestry and land-use change in tropical Asia. Global Change Biology 5 (4): 481–492.

Howe HF, Miriti MN, 2000. No question: seed dispersal matters. Trends in Ecology & Evolution 15 (11): 434–436.

Huang W, Pohjonen V, Johansson S, Nashanda M, Katigula MI, Luukkanen O, 2003. Species diversity, forest structure and species composition in Tanzanian tropical forests. Forest Ecology and Management 173 (1-3): 11–24.

Hung PQ, 2008. Structure and light factor in differently logged moist forests in Huong Son-Vu Quang, Vietnam. Göttingen University, Germany.

Indarto J, Kaneko S, Kawata K, 2015. Do forest permits cause deforestation in Indonesia? International Forestry Review 17 (2): 165–181.

Isango JA, 2007. Stand Structure and Tree Species Composition of Tanzania Miombo Woodlands: A Case Study from Miombo Woodlands of Community Based Forest Management in Iringa District. In, Management of indigenous tree species for ecosystem restoration and wood production in semi-arid Miombo woodland in Eastern Africa, Morogor, Tanzania Working Papers of the Finnish Research Institute 50: 43-56.

ITTO, 2006. Status of tropical forest management 2005. International Tropical Timber Organization, Yokohama, Japan.

Johns RJ, 1985. The vegetation and flora of the south Naru area (Madang Province), Papua New Guinea. Klinkii 3 (4): 70–83.

Johns RJ, 1995. Endemism in the Malesian flora. Curtis's Botanical Magazine 12 (2): 95–110.

Jones EW, 1950. Some aspects of natural regeneration in the Benin rain forest. Empire Forestry Review 108–124.

Jones EW, 1955. Ecological studies on the rain forest of southern Nigeria: IV. The plateau forest of the Okomu Forest Reserve. Journal of Ecology 43 (2): 564–594.

Joppa LN, Loarie SR, Pimm SL, 2009. On population growth near protected areas. PloS One 4 (1): e4279.

Kammesheidt L, Köhler P, Huth A, 2001. Sustainable timber harvesting in Venezuela: a modelling approach. Journal of Applied Ecology 38 (4): 756–770.

Kartodihardjo H, Supriono A, 2000. The impact of sectoral development on natural forest conversion and degradation: The case of timber and tree crop plantations in Indonesia. Center for International Forestry Research, Bogor, Indonesia.

Kerstiens G, 2001. Meta-analysis of the interaction between shade-tolerance, light environment and growth response of woody species to elevated CO2. Acta Oecologica 22 (1): 61–69.

Khaing N, 2013. Structure and Site Conditions of Dry Deciduous Forests in Central Myanmar. Georg-August-Universität Göttingen, Göttingen,. 168 pp.

Kiapranis R, 1991. Plant species enumeration in a lowland rain forest in Papua New Guinea. Research on Multipurpose Tree Species in Asia 98–101.

Kimmins JP, 2004. Forest Ecology: A foundation for sustainable forest management and environmental ethics in forestry, rd Prentice Hall. Upper Saddle River, NJ, USA.

King DA, 1996. Allometry and life history of tropical trees. Journal of Tropical Ecology 25–44.

Klinge H, Rodrigues WA, Brunig E, Fittkau EJ (eds), 1975. Biomass and structure in a central Amazonian rain forest. Tropical Ecological Systems. Springer, Berlin. pp. 115-122.

Koba K, Hirobe M, Koyama L, Kohzu A, Tokuchi N, Nadelhoffer KJ, Wada E, Takeda H. 2003. Natural 15 N abundance of plants and soil N in a temperate coniferous forest. Ecosystems 6 (5): 457–469.

Kohyama T, Hara T, Tadaki Y, 1990. Patterns of trunk diameter, tree height and crown depth in crowded Abies stands. Annals of Botany 65 (5): 567–574.

Koopmans CJ, van Dam D, Tietema A, Verstraten JM, 1997. Natural 15 N abundance in two nitrogen saturated forest ecosystems. Oecologia 111 (4): 470–480.

Kostermans ,AJGH, 1957. Lauraceae. Reinwardtia 4 (2): 193–256.

Kukkonen M, Hohnwald S, 2009. Comparing floristic composition in treefall gaps of certified, conventionally managed and natural forests of northern Honduras. Annals of Forest Science 66 (8): 809.

Kurokawa H, Yoshida T, Nakamura T, Lai J, Nakashizuka T, 2003. The age of tropical rain-forest canopy species, Borneo ironwood (Eusideroxylon zwageri), determined by 14C dating. Journal of Tropical Ecology 19 (1): 1–7.

Kuswandi R, Remetwa H, Tokede MJ, 1993. Analisis komposisi jenis vegetasi di kelompok hutan Prafi dan Tuanwouwi, Manokwari. Paratropika 1 (2): 19–30.

Kuswandi R, 2010. Metode pengaturan hasil hutan alam bekas tebangan melalui pendekatan model dinamika sistem di Kabupaten Boven Digul, Papua. Thesis: Universitas Gadjah Mada, Yogyakarta-Indonesia.

Kuswandi R, 2017. Model pertumbuhan tegakan hutan alam bekas tebangan dengan sistem tebang pilih di Papua. Jurnal Pemuliaan Tanaman Hutan 11 (1): 45–56.

Kuusipalo J, Hadengganan S, Ådjers G, Sagala APS, 1997. Effect of gap liberation on the performance and growth of dipterocarp trees in a logged-over rainforest. Forest Ecology and Management 92 (1-3): 209–219.

Lamprecht H, 1989. Silviculture in the tropics: tropical forest ecosystems and their tree species: possibilities and methods for their long-term utilization. Technical Cooperation-Federal Republic of Germany (GTZ), Eschborn.

Langner A, Miettinen J, Siegert F, 2007. Land cover change 2002–2005 in Borneo and the role of fire derived from MODIS imagery. Global Change Biology 13 (11): 2329–2340.

Laporte NT, Stabach JA, Grosch R, Lin TS, Goetz SJ, 2007. Expansion of industrial logging in Central Africa. Science 316 (5830): 1451.

Larjavaara M, Muller-Landau HC, 2013. Measuring tree height: a quantitative comparison of two common field methods in a moist tropical forest. Methods in Ecology and Evolution 4 (9): 793–801.

Laurance WF, Albernaz AKM, Schroth G, Fearnside PM, Bergen S, Venticinque EM, Da Costa C, 2002. Predictors of deforestation in the Brazilian Amazon. Journal of Biogeography 29 (5-6): 737–748.

Lee DK, 2009. Challenging forestry issues in Asia and their strategies. In: The future of forests in Asia and the Pacific. Proceedings of International Conference on the Outlook for Asia-Pacific Forests to 2020, Sess. 2, Chiang Mai, 16-18 Oct 2007. FAO, Bangkok, Thailand. 65.

Leenhouts PW, Kalkman C, Lam HJ, 1955. Burseraceae. Flora Malesiana-Series 1, Spermatophyta 5 (1): 209–296.

Lieffers VJ, Messier C, Stadt KJ, Gendron F, Comeau PG, 1999. Predicting and managing light in the understory of boreal forests. Canadian Journal of Forest Research 29 (6): 796–811.

Loetsch F, Haller KE, Zöhrer F, 1973. Forest inventory. BLV Verlagsgesellschaft, München.

Magurran AE, 1988. Ecological diversity and its measurement. Springer Science & Business Media, Dordrecht, The Netherlands.

Marwa J, 2009. Model Dinamik Pengaturan Hasil Hutan Tidak Seumur dan Kontribusinya Terhadap Ekonomi Daerah (Studi Kasus IUPHHK P. Bina Balantak Utama Kabupaten Sarmi, Papua). Thesis unpublished: Bogor Agricultural University, Bogor-Indonesia.

Mason WL, Edwards C, Hale SE, 2004. Survival and early seedling growth of conifers with different shade tolerance in a Sitka spruce spacing trial and relationship to understorey light climate. Silva Fennica 38 (4): 357–370.

Mawazin M, Subiakto A, 2013. Keanekaragaman dan komposisi jenis permudaan alam hutan rawa gambut bekas tebangan di Riau (species diversity and composition of logged over peat swamp forest in Riau). Indonesian Forest Rehabilitation Journal 1 (1): 59–73.

McCune B, Grace JB, Urban DL, 2002. Analysis of Ecological Communities; MjM Software Design: Gleneden Beach, OR, USA, 2002. Google Scholar.

Meyer HA, 1953. Forest mensuration. Penns Velley Publishing, Inc., State College, Pa. USA. 357 pp.

Miettinen J, Shi C, Liewl SC, 2011. Deforestation rates in insular Southeast Asia between 2000 and 2010. Global Change Biology 17 (7): 2261–2270.

Miettinen J, Hooijer A, Vernimmen R, Liew SC, Page SE, 2017. From carbon sink to carbon source: extensive peat oxidation in insular Southeast Asia since 1990. Environmental Research Letters 12 (2): 24014.

Mitlöhner R, 1998. Pflanzeninterne Potentiale als Indikatoren für den tropischen Standort. Shaker Verlag, Aachen, Germany. 238 pp.

MoEFOR, 2017. Statistik Lingkungan Hidup dan Kehutanan (Environment and Forestry Statistics). Kementrian Lingkungan Hidup dan Kehutanan Indonesia/Indonesian Ministry of Environment and Forestry (MoEFOR), Jakarta-Indonesia.

MoFOR, 2008. Statistik Kehutanan Indonesia (Indonesian Forestry Statistics): 1. Planologi Kehutanan (Forestry Planning). Kementrian Kehutanan Indonesia/Indonesian Ministry of Forestry (MoFOR), Jakarta-Indonesia.

Mori SA, BooM BM, Carvalho AM de, dos Santos TS, 1983. Southern Bahian moist forests. The Botanical Review 49 (2): 155–232.

Nasi R, Frost PGH, 2009. Sustainable forest management in the tropics: is everything in order but the patient still dying? Ecology and Society 14 (2).

Nawir AA, Rumboko L, 2007. History and state of deforestation and land degradation. Center for International Forestry Research, Jakarta, Indonesia, Bogor, Indonesia.

Newman J, 2008. Last Frontier: Illegal Logging in Papua and China's Massive Timber Theft. DIANE Publishing/Telapak/Environmental Investigation Agency, Bogor, Indonesia.

Niklas KJ, 1995. Size-dependent allometry of tree height, diameter and trunk-taper. Annals of Botany 75 (3): 217–227.

Nyland RD, 2016. Silviculture: concepts and applications. Waveland Press.

Oatham M, Beehler BM, 1995. Richness, taxonomic composition, and species patchiness in three lowland forest plots in Papua New Guinea. Forest Biodiversity Research, Monitoring and Modeling-conceptual Background and Old World Case Studies 20: 613–631.

O'Brien ST, Hubbell SP, Spiro P, Condit R, Foster RB, 1995. Diameter, height, crown, and age relationship in eight neotropical tree species. Ecology 76 (6): 1926–1939.

Ogawa H, 1965. Comparative ecological studies on three main types of forest vegetation in Thailand. II. Plant biomass. Nature and Life in Southeast Asia 4: 49–80.

Oliver CD, 1992. Achieving and maintaining biodiversity and economic productivity. Journal of Forestry 90 (9): 20–25.

Oliver CD, Larson BC, 1990. Forest stand dynamics. Biological resource management series. McGraw-Hill, New York.

Page SE, Rieley JO, Banks CJ, 2011. Global and regional importance of the tropical peatland carbon pool. Global Change Biology 17 (2): 798–818.

Paijmans K, 1970. An analysis of four tropical rain forest sites in New Guinea. Journal of Ecology 77–101.

Paijmans K, 1976. New Guinea vegetation. Commonwealth Scientific and Industrial Research Organization in association with the Australian National University Press, Canberra, ACT, Australia.

Pakwel A, Ryhage R, Wickman FE, 1957. Natural variations in the relative abundances of the nitrogen isotopes. Geochimica et Cosmochimica Acta 11 (3): 165–170.

Park A, Justiniano MJ, Fredericksen TS, 2005. Natural regeneration and environmental relationships of tree species in logging gaps in a Bolivian tropical forest. Forest Ecology and Management 217 (2-3): 147–157.

Pate JS, 1973. Uptake, assimilation and transport of nitrogen compounds by plants. Soil Biology and Biochemistry 5 (1): 109–119.

Peña-Claros M, Fredericksen TS, Alarcón A, Blate GM, Choque U, Leaño C, Licona JC, Mostacedo B, Pariona W, Villegas Z, 2008. Beyond reduced-impact logging: silvicultural treatments to increase growth rates of tropical trees. Forest Ecology and Management 256 (7): 1458–1467.

Petocz RG, 1989. Conservation and development in Irian Jaya: a strategy for rational resource utilization. E.J. Brill, Leiden, The Netherlands.

Pfaff A, Robalino J, Walker R, Aldrich S, Caldas M, Reis E, Perz S, Bohrer C, Arima E, Laurance W, 2007. Road investments, spatial spillovers, and deforestation in the Brazilian Amazon. Journal of Regional Science 47 (1): 109–123.

Pham MT, 2012. Structure and Regeneration of Lowland Tropical Moist Evergreen Forests in North and Central Vietnam. In Dissertation. Georg-August-Universität Göttingen, Göttingen,

Philip MS, 1994. Measuring trees and forests. CAB international, University of Michigan, USA.

Pires JM, Prance GT (eds), 1977. The Amazon Forest: a natural heritage to be preserved. In Conference Proceding Extinction is Forever 11-13 May 1976, Bronx, NY, USA.

Pitman NCA, Terborgh J, Silman MR, Nuñez V P, 1999. Tree species distributions in an upper Amazonian forest. Ecology 80 (8): 2651–2661.

Polhemus DA, 2007. Tectonic Geology of Papua. In A.J. Marshall and B.M. Beehler (Ed.). The Ecology of Papua Part I, The Ecology of Indonesia series, Vol. IV. 137–164.

Polhemus DA, Allen GR, 2007. Freshwater biogeography of Papua. In A.J. Marshall and B.M. Beehler (Ed.). The Ecology of Papua Part I, The Ecology of Indonesia series, Vol. IV. 207–245.

Poorter L, 1999. Growth responses of 15 rain-forest tree species to a light gradient: the relative importance of morphological and physiological traits. Functional Ecology 13 (3): 396–410.

Poorter L, Werger MJA, 1999. Light environment, sapling architecture, and leaf display in six rain forest tree species. American Journal of Botany 86 (10): 1464–1473.

Poorter L, Bongers F, 2006. Leaf traits are good predictors of plant performance across 53 rain forest species. Ecology 87 (7): 1733–1743.

Poorter L, Bongers F, van Rompaey R, Klerk M de, 1996. Regeneration of canopy tree species at five sites in West African moist forest. Forest Ecology and Management 84 (1): 61–70.

Poorter L, Hawthorne W, Bongers F, Sheil D, 2008. Maximum size distributions in tropical forest communities: relationships with rainfall and disturbance. Journal of Ecology 495–504.

Poulson SR, Chamberlain CP, Friedland AJ, 1995. Nitrogen isotope variation of tree rings as a potential indicator of environmental change. Chemical Geology 125 (3-4): 307–315.

Primack RB, Hall P, 1992. Biodiversity and forest change in Malaysian Borneo. BioScience 42 (11): 829–837.

PT. TTL, 2011. PT. Tunas Timber Lestari (TTL) Company Profile. PT. Tunas Timber Lestari (TTL), Jakarta-Indonesia.

Puslitbangtanak, 2000. Atlas Sumber Daya Tanah Eksplorasi Indonesia. Skala 1:1.000.000. Pusat Penelitian dan Pengembangan Tanah dan Agroklimat (Puslitbangtanak), Bogor-Indonesia.

Putz FE, Sist P, Fredericksen T, Dykstra D, 2008. Reduced-impact logging: challenges and opportunities. Forest Ecology and Management 256 (7): 1427–1433.

Putz FE, Zuidema PA, Synnott T, Peña-Claros M, Pinard MA, Sheil D, Vanclay JK, Sist P, Gourlet-Fleury S, Griscom B, 2012. Sustaining conservation values in selectively logged tropical forests: the attained and the attainable. Conservation Letters 5 (4): 296–303.

Pyke CR, Condit R, Aguilar S, Lao S, 2001. Floristic composition across a climatic gradient in a neotropical lowland forest. Journal of Vegetation Science 12 (4): 553–566.

Rai SN, 1979. Diameter/height, and diameter/girth relationship of some rain forest tree species of Karnataka--India. Malaysian Forester 42 (1): 53–58.

RePPProT, 1987. Review of Phase I Results: Irian Jaya. Regional Physical Planning Programme for Transmigration, ODA, for Gov. Indonesia. Ministry of Transmigration, Jakarta, Indonesia. 2 vol.+ 3 series of 1: 250,000 scale maps.

Rich PM, Helenurm K, Kearns D, Morse SR, Palmer MW, Short L, 1986. Height and stem diameter relationships for dicotyledonous trees and arborescent palms of Costa Rican tropical wet forest. Bulletin of the Torrey Botanical Club 241–246.

Richards PW, 1996. The tropical rain forest: an ecological study. ed. 2. Cambridge University Press, Cambridge, UK.

RKU PT. TTL, 2009. Rencana Kerja Usaha (RKU) Pemanfaatan Hasil Hutan Kayu dalam Hutan Alam (PHHK-HA) pada Hutan Produksi Periode 2009-2018. PT. Tunas Timber Lestari (TTL), Boven Digul, Papua, Indonesia.

RKU PT. WMT, 2012. Rencana Kerja Usaha (RKU) Pemanfaatan Hasil Hutan Kayu dalam Hutan Alam (PHHK-HA) pada Hutan Produksi Periode 2012-2021. PT. Wapoga Mutiara Timber (WMT) Unit II, Sarmi, Papua, Indonesia.

Robinson D, Handley LL, Scrimgeour CM, Gordon DC, Forster BP, Ellis RP, 2000. Using stable isotope natural abundances (δ 15 N and δ 13 C) to integrate the stress responses of wild barley (Hordeum spontaneum C. Koch.) genotypes. Journal of Experimental Botany 51 (342): 41–50.

Rollet B, 1974. L'architecture des forêts denses humides sempervirentes de plaines. Centre Technique Forestier Tropical, Nogent-sur-Marne, France.

Sabogal C, 1992. Regeneration of tropical dry forests in Central America, with examples from Nicaragua. Journal of Vegetation Science 3 (3): 407–416.

Sadono R, 2014. Structural equation model for population of Merbau (Intsia bijuga, Colebr. O. Kuntze) in Gunung Meja Natural Tourism Park of Manokwari in Papua. Advances in Environmental Biology 8 (17): 1012–1019.

Schulz JP, 1960. Ecological studies on rain forest in northern Suriname. Mededelingen van het Botanisch Museum en Herbarium van de Rijksuniversiteit te Utrecht 163 (1): 1–267.

Schulze M, 2008. Technical and financial analysis of enrichment planting in logging gaps as a potential component of forest management in the eastern Amazon. Forest Ecology and Management 255 (3-4): 866–879.

Schulze M, Grogan J, Landis RM, Vidal E, 2008. How rare is too rare to harvest?: management challenges posed by timber species occurring at low densities in the Brazilian Amazon. Forest Ecology and Management 256 (7): 1443–1457.

Simpson EH, 1949. Measurement of diversity. Nature 163 (4148): 688.

Sist P, Nguyen-Thé N, 2002. Logging damage and the subsequent dynamics of a dipterocarp forest in East Kalimantan (1990–1996). Forest Ecology and Management 165 (1-3): 85–103.

Sist P, Nolan T, Bertault J-G, Dykstra D, 1998. Harvesting intensity versus sustainability in Indonesia. Forest Ecology and Management 108 (3): 251–260.

Sist P, Sheil D, Kartawinata K, Priyadi H, 2003a. Reduced-impact logging in Indonesian Borneo: some results confirming the need for new silvicultural prescriptions. Forest Ecology and Management 179 (1-3): 415–427.

Sist P, Fimbel R, Sheil D, Nasi R, Chevallier M-H, 2003b. Towards sustainable management of mixed dipterocarp forests of South-east Asia: moving beyond minimum diameter cutting limits. Environmental Conservation 30 (4): 364–374.

Smith DM, 1986. The practice of silviculture. John Wiley & Sons, inc, New York, USA.

Sodhi NS, Posa MRC, Lee TM, Bickford D, Koh LP, Brook BW, 2010. The state and conservation of Southeast Asian biodiversity. Biodiversity and Conservation 19 (2): 317–328.

Spies TA, 1998. Forest structure: a key to the ecosystem. Northwest Science 72: 34–36.

Stibig H-J, Achard F, Carboni S, Raši R, Miettinen J, 2014. Change in tropical forest cover of Southeast Asia from 1990 to 2010. Biogeosciences 11 (2): 247–258.

Struhsaker TT, 1997. Ecology of an African rain forest: logging in Kibale and the conflict between conservation and exploitation. University Press of Florida, Gainesville, USA.

Sumida A, Ito H, Isagi Y, 1997. Trade-off between height growth and stem diameter growth for an evergreen Oak, Quercus glauca, in a mixed hardwood forest. Functional Ecology 11 (3): 300–309.

Takagi K, Kotsuka C, Fukuzawa K, Kayama M, Makoto K, Watanabe T, Nomura M, Fukazawa T, Takahashi H, Hojyo H, 2010. Allometric relationships and carbon and nitrogen contents for three major tree species (Quercus crispula, Betula ermanii, and Abies sachalinensis) in Northern Hokkaido, Japan. Eurasian Journal of Forest Research 13 (1): 1–7.

Takeuchi WN, 2007. Introduction to the flora of Papua. In A.J. Marshall and B.M. Beehler (Ed.). The Ecology of Papua Part I, The Ecology of Indonesia series, Vol. IV. 269–279.

Teketay D, 1995. Floristic composition of Dakata Valley, southeast Ethiopia: An implication for the conservation of biodiversity. Mountain Research and Development 15 (2): 183–186.

Teketay D, Bekele T, 1995. Floristic composition of Wof-Washa natural forest, Central Ethiopia: Implications for the conservation of biodiversity. Feddes Repertorium 106 (1-5): 127–147.

ter Steege H, Welch I, Zagt R, 2002. Long-term effect of timber harvesting in the Bartica Triangle, Central Guyana. Forest Ecology and Management 170 (1-3): 127–144.

Thaman RR, Thomson LAJ, DeMeo R, Areki F, Elevitch CR, 2006. Intsia bijuga (vesi). Species Profiles for Pacific Island Agroforestry.

Tilman D, 2020. Plant Strategies and the Dynamics and Structure of Plant Communities.(MPB-26). Monographs in Population Biology: Vol. 26. Princeton University Press, Chichester, UK.

Toledo M, Poorter L, Peña-Claros M, Alarcón A, Balcázar J, Chuviña J, Leaño C, Licona JC, ter Steege H, Bongers F, 2011. Patterns and determinants of floristic variation across lowland forests of Bolivia. Biotropica 43 (4): 405–413.

Tuomisto H, Ruokolainen K, Yli-Halla M, 2003. Dispersal, environment, and floristic variation of western Amazonian forests. Science 299 (5604): 241–244.

Ullrich WR, 1992. Transport of nitrate and ammonium through plant membranes. Nitrogen Metabolism of Plants 121–137.

UNFCCC, 2008. Report of the Conference of the Parties on its thirteenth session, held in Bali from 3 to 15 December 2007. Addendum Part Two: Action taken by the Conference of the Parties at its thirteenth session. Bali, Indonesia.

Valle D, Phillips P, Vidal E, Schulze M, Grogan J, Sales M, van Gardingen P, 2007. Adaptation of a spatially explicit individual tree-based growth and yield model and long-term comparison between reduced-impact and conventional logging in eastern Amazonia, Brazil. Forest Ecology and Management 243 (2-3): 187–198.

van Gardingen PR, Valle D, Thompson I, 2006. Evaluation of yield regulation options for primary forest in Tapajos National Forest, Brazil. Forest Ecology and Management 231 (1-3): 184–195.

van Laar A, Akça A, 2007. Forest mensuration. Springer Science & Business Media, Dordrecht, The Netherlands.

Vanclay JK, 1991. Aggregating tree species to develop diameter increment equations for tropical rainforests. Forest Ecology and Management 42 (3-4): 143–168.

Vanclay JK, 1994. Modelling forest growth and yield: applications to mixed tropical forests. CAB international, Wallingford, UK.

Verissimo A, Barreto P, Tarifa R, Uhl C, 1995. Extraction of a high-value natural resource in Amazonia: the case of mahogany. Forest Ecology and Management 72 (1): 39–60.

Vieira S, Trumbore S, Camargo PB, Selhorst D, Chambers JQ, Higuchi N, Martinelli LA, 2005. Slow growth rates of Amazonian trees: consequences for carbon cycling. Proceedings of the National Academy of Sciences 102 (51): 18502–18507.

Villegas Z, Peña-Claros M, Mostacedo B, Alarcón A, Licona JC, Leaño C, Pariona W, Choque U, 2009. Silvicultural treatments enhance growth rates of future crop trees in a tropical dry forest. Forest Ecology and Management 258 (6): 971–977.

Virginia RA, Delwiche CC, 1982. Natural 15 N abundance of presumed N 2-fixing and non-N 2-fixing plants from selected ecosystems. Oecologia 54 (3): 317–325.

Wagner B, 2016. Structural Analysis of Tropical Lowland Rainforests in District Sausapor of Tambrauw Regency in West Papua, Indonesia. In, Master Thesis. Georg – August – University Göttingen, Georg – August – University Göttingen, Göttingen.

Walker D, Hope GS, 1982. Late Quaternary vegetation history. In J.L. Gressitt (Ed.). Biogeography and ecology of New Guinea, Monographiae Biologiacae. Dr. W. Junk Publisher, The Haque. p 263-285.

Wallsgrove RM, Lea PJ, Miflin BJ, 1979. Distribution of the enzymes of nitrogen assimilation within the pea leaf cell. Plant Physiology 63 (2): 232–236.

Walter H, Lieth H, 1967. Klimadiagramm - Weltatlas. Veb Gustav Fischer Verlag, Jena.

Warner RL, Huffaker RC, 1989. Nitrate transport is independent of NADH and NAD (P) H nitrate reductases in barley seedlings. Plant Physiology 91 (3): 947–953.

Weidelt HJ, 1998. Tropical Silviculture-Part II, Lecture notes from institute of Tropical Silviculture, University of Göttingen, Germany. Unpublished.

Weiner J, Thomas SC, 1992. Competition and allometry in three species of annual plants. Ecology 73 (2): 648–656.

Whitmore TC, 1984. Tropical rain forests of the Par East. Oxford. Clarendon Press.

Whitmore T, 1989. Canopy gaps and the two major groups of forest trees. Ecology 70 (3): 536–538.

Wilcove DS, Giam X, Edwards DP, Fisher B, Koh LP, 2013. Navjot's nightmare revisited: logging, agriculture, and biodiversity in Southeast Asia. Trends in Ecology & Evolution 28 (9): 531–540.

Wilde WJJO de, 2000. Myristicaceae. Flora Malesiana-Series 1, Spermatophyta 14 (1): 1–632.

Wilson KA, Meijaard E, Drummond S, Grantham HS, Boitani L, Catullo G, Christie L, Dennis R, Dutton I, Falcucci A, 2010. Conserving biodiversity in production landscapes. Ecological Applications 20 (6): 1721–1732.

Womersley JS, Henty EE, Conn BJ, 1978. Handbooks of the Flora of Papua New Guinea. Melbourne University Press, Carlton, Vic. 3 Bände.

Wright DD, Jessen JH, Burke P, Silva Garza HG de, 1997. Tree and Liana Enumeration and Diversity on a One-Hectare Plot in Papua New Guinea 1. Biotropica 29 (3): 250–260.

Yeom FBC, Chandrasekharan C, 2001. Achieving sustainable forest management in Indonesia. ITTO Newsletter.

Yoneyama T, Kamachi K, Yamaya T, Mae T, 1993. Fractionation of nitrogen isotopes by glutamine synthetase isolated from spinach leaves. Plant and Cell Physiology 34 (3): 489–491.

Yoneyama T, Omata T, Nakata S, Yazaki J, 1991. Fractionation of nitrogen isotopes during the uptake and assimilation of ammonia by plants. Plant and Cell Physiology 32 (8): 1211–1217.

Zimmerman BL, Kormos CF, 2012. Prospects for sustainable logging in tropical forests. BioScience 62 (5): 479–487.

Zöhrer F, 1980. Forstinventur. Ein Leitfaden für Studium und Praxis. Pareys Studientexte 26. Parey, Hamburg.

Appendices

Appendix 1. Species Important Value Indices (IVI) in the intact forest of Tunas and Bonggo

No.	Species Name	Bonggo Forest		Tunas Forest	
		Flat	Steep	Flat	Steep
		IVI (%)			
1	*Actinodaphne nitida* Teschner			2.97	2.71
2	*Adina multifolia* Havil.			1.44	0.98
3	*Aglaia* sp	0.79	1.18	5.60	5.01
4	*Albizia falcataria* (L.) Fosberg	0.56	1.30		
5	*Alphitonia macrocarpa* Mansf.	0.93	1.67		
6	*Alstonia scholaris* R.Br.	2.41	4.22		
7	*Alstonia spectabilis* R.Br.				0.92
8	*Anisoptera polyandra* Blume	16.35	16.55	10.02	5.46
9	*Anisoptera thurifera* Blume		1.54	1.57	2.01
10	*Anthocephalus cadamba*	1.64	0.29		
11	*Aquilaria polyantha*				0.23
12	*Aquilaria* sp	0.99	0.44		
13	*Arthocarpus* spp.	2.88	4.75		
14	*Baccaurea* sp				0.36
15	*Beilschmiedia* sp			8.07	4.68
16	*Blumeodendron amboinicum*			10.41	9.38
17	*Buchanania macrocarpa* Lauterb.	1.02	0.78		
18	*Calophyllum costatum* F.M. Bailey			2.26	4.04
19	*Calophyllum inophyllum* Lauterb.	1.85	2.54	2.86	1.32
20	*Calophyllum papuanum* Lauterb.	3.72	6.52	1.31	0.71
21	*Campnosperma auriculatum* (Blm) Hook.f.	6.70	1.74	0.74	0.32
22	*Cananga odorata* Hook. f. & Thomson	5.57	3.43		
23	*Canarium asperum* Benth.	4.74	8.11	8.41	11.59

No.	Species Name	Bonggo Forest		Tunas Forest	
		Flat	Steep	Flat	Steep
		IVI (%)			
24	*Canarium australianum* F.Muell.	2.04	0.90		
25	*Canarium hirsutum* Willd.	0.72		2.07	2.47
26	*Canarium indicum* L.	8.08	3.93	2.71	4.99
27	*Canarium molucense*		0.99		
28	*Canarium rigidum*	1.58	4.26		
29	*Canarium* sp	4.53	3.19		2.66
30	*Carallia brachiata* (Lour.) Merr.			4.11	3.59
31	*Carallia* sp.			2.61	1.06
32	*Celtis latifolia* Planch.			0.29	
33	*Celtis rigescens* Planch.	17.91	17.52		
34	*Chisocheton ceramicus* C. DC.	2.80	3.96	4.93	6.20
35	*Chrysophyllum* sp			0.27	
36	*Cinnamomum culilawan* Blume	1.06	1.46	1.91	0.77
37	*Crudia papuana* Kosterm.	0.37			
38	*Cryptocarya palmerensis* C.K. Allen			0.65	1.72
39	*Dacryodes* sp			4.28	8.74
40	*Dillenia alata* A. DC.	1.10	2.87		
41	*Dillinea ovalifolia*				1.12
42	*Diospyros* sp	2.40	2.40	0.59	1.39
43	*Dracontomelum edule* Merr.	4.13	1.58		
44	*Drypetes globosa* (Merr.) Pax & K. Hoffm.			4.55	4.54
45	*Duabanga moluccana* Blume	4.84	1.17		
46	*Dysoxylum acutangulum* Miq.			2.67	0.97
47	*Dysoxylum amooroides* Miq.	1.11	1.26		
48	*Dysoxylum mollissimum* Blume			6.13	6.32
49	*Dysoxylum octandrum* Merr.			0.31	1.02

No.	Species Name	Bonggo Forest		Tunas Forest	
		Flat	Steep	Flat	Steep
		IVI (%)			
50	*Elaeocarpus angustifolius* Blume			2.36	4.85
51	*Elaeocarpus sphaericus* Schum.	0.86	0.70		
52	*Endospermum medullosum* L.S. Sm.	7.31	1.78		0.26
53	*Eucalyptus deglupta* Blume	0.73	1.32		
54	*Euodia elleryana* F. Muell.			0.83	0.80
55	*Fagraea* sp			1.21	1.73
56	*Ficus variegata* Blume		0.42		
57	*Ganophyllum falcatum* Blume				1.13
58	*Garcinia celebica* L.			2.43	6.34
59	*Garcinia dulcis* (Roxb.) Kurz	0.67	0.74	0.88	1.20
60	*Garuga floribunda* Decne.			0.88	0.22
61	*Gironniera subaequalis* Plancho			7.26	5.91
62	*Gnetum gnemon* L.	0.99	2.05	1.63	0.24
63	*Gonocaryum tiriform*			0.62	1.26
64	*Haplolobus floribundus* H.J.Lam		1.01	5.94	4.67
65	*Haplolobus lanceolatus* H.J.Lam			0.57	0.36
66	*Homalium foetidum*	23.52	14.32	1.12	1.14
67	*Hopea papuana* Diels		2.25	8.38	8.23
68	*Horsfieldia irya* Warb.			0.50	
69	*Horsfieldia sylvestris* Warb.	0.35	1.03	1.14	
70	*Inocarpus* sp	1.05	0.22		
71	*Intsia bijuga* Kuntze	22.67	23.88		
72	*Kayea coriacea* Stevens				0.33
73	*Kingiodendron novoguineense* Verdc.				0.22
74	*Koordersiodendron papuanum*		1.98		
75	*Linociera macrophylla* Wall.			5.43	5.66

No.	Species Name	Bonggo Forest		Tunas Forest	
		Flat	Steep	Flat	Steep
		IVI (%)			
76	*Linociera oblongifolia* Koord.	1.07	0.47		
77	*Lithocarpus rufovillosus* Rehder			5.38	5.22
78	*Litsea firma*			0.26	
79	*Litsea ledermannii* Teschner	11.37	12.60	3.44	2.52
80	*Litsea timoriana* Span.			16.33	14.47
81	*Litsea tuberculata* Boerl.				2.62
82	*Litsea irianensis* Kosterm.				1.07
83	*Lophopetalum* spp	1.16			
84	*Macaranga beillei* Prain	0.67	2.63		
85	*Mangifera* sp	1.03	0.79		
86	*Manilkara fasciculata* H.J. Lam & Maas			5.66	3.24
87	*Manilkara* sp		0.43		
88	*Maniltoa plurijuga*			1.67	1.55
89	*Mastixiodendron pachyclados* Melch.	2.92	6.11		
90	*Mastixiodendron plectocarpum* S.P.Darwin			0.77	2.11
91	*Medusanthera polot*			2.34	8.11
92	*Myristica argentea*			2.41	0.72
93	*Myristica fatua* Houtt.	0.64	1.09	2.47	1.32
94	*Myristica globosa* W.J. de Wilde			2.26	2.60
95	*Myristica hollrungii* Warb.			0.96	0.32
96	*Myristica lepidota* Blume	1.04	3.03	0.51	
97	*Myristica* sp	1.68	3.03		
98	*Myristica sulcata* Warb.	9.46	11.90	0.80	4.68
99	*Myristica tubiflora* Blume	15.00	7.09	4.97	4.88
100	*Nauclea orientalis*		1.53		
101	*Neoscortechinia* sp			0.64	1.04

No.	Species Name	Bonggo Forest		Tunas Forest	
		Flat	Steep	Flat	Steep
		IVI (%)			
102	*Palaquium galactoxylum* H.J.Lam			0.79	1.86
103	*Palaquium lobbianum* Burck			1.49	2.74
104	*Palaquium obtusifolium* Burck	7.06	7.56	0.43	0.82
105	*Parartocarpus* spp	3.94	2.69		
106	*Parastemon versteeghii* Merr. & L.M.Perry			0.25	
107	*Pimelodendron amboinicum* Hassk.	9.82	8.46	8.69	6.46
108	*Planchonella anteridifera* H.J. Lam			4.16	3.59
109	*Planchonella firma* (Miq.) Dubard			5.37	2.89
110	*Planchonella* sp			1.25	
111	*Podocarpus amara*				0.25
112	*Podocarpus nerlifolius* D.Do	0.61		1.40	
113	*Pometia acuminata*	3.03	6.26	4.33	3.64
114	*Pometia pinnata* J.R. Forst. & G. Forst.	15.19	14.07	2.04	1.97
115	*Prunus javanica* Miq.			0.61	1.51
116	*Pterocarpus indicus* Willd.	8.05	6.04		
117	*Pterygota horsfieldii* Kosterm.	5.92	6.11		
118	*Reinwardtiodendron celebicum* Koord.			0.54	
119	*Saccopetalum* sp				0.57
120	*Semercarpus anacardium*			0.26	
121	*Sloanea pullei* A.C. Sm.	0.35	0.31	1.02	2.23
122	*Spathiostemon javensis* Blume.			2.04	2.12
123	*Spondias dulcis* G. Forst.	1.53	1.96		
124	*Stemonurus javanicum*			0.63	0.22
125	*Sterculia parkinsonii*	0.36	1.17		
126	*Sterculia* sp			0.40	1.04
127	*Swietenia macrophylla* King		1.65		

No.	Species Name	Bonggo Forest		Tunas Forest	
		Flat	Steep	Flat	Steep
		IVI (%)			
128	*Syzygium anomalum* Lauterb.	8.70	11.53	5.99	6.98
129	*Syzygium gustavioides* B.Hyland			5.82	3.53
130	*Syzygium hylophilum* Merr. & L.M.Perry	0.39	2.08	4.91	4.61
131	*Syzygium* sp1	11.64	6.90		
132	*Syzygium* sp2			1.58	2.12
133	*Syzygium versteegii* Merr. & L.M.Perry	1.71	4.23	7.28	7.69
134	*Teijsmanniodendron bogoriense* Koord.			9.89	6.78
135	*Terminalia arborea* Koord.	3.17	6.59		
136	*Terminalia* sp				0.22
137	*Timonius belensis* Merr. & L.M.Perry	0.34			
138	*Toona sureni* B. (Meril)	6.90	5.36		
139	*Vatica rassak* Blume	3.72	1.48	35.80	31.93
140	*Vitex pubescens* (L.) Vahl			1.20	1.76
141	*Vitex* sp	0.59	2.57		
142	*Whiteodendron papuana*			3.15	1.47
143	*Xanthostemon crenulatus* C.T. White			2.86	2.51
144	*Zizyphus* sp				0.22

Appendix 2. Species Important Value Indices (IVI) in logged forests of Tunas and Bonggo

No.	Species Name	Bonggo Forest		Tunas Forest	
		4-year logged	8-year logged	4-year logged	8-year logged
		IVI (%)			
1	*Aceratium diversifolium*			0.91	2.40
2	*Actinodaphne nitida* Teschner				2.99
3	*Adenanthera microsperma* Teijsm. & Binn.	2.02	0.84		
4	*Adina multifolia* Havil.			5.43	2.50
5	*Aglaia* sp	1.61	0.61	0.90	1.33
6	*Albizia falcataria* (L.) Fosberg	1.65			
7	*Alphitonia incana* (Roxb.) Kurz				0.94
8	*Alphitonia macrocarpa* Mansf.		2.29		
9	*Alstonia scholaris* R.Br.		2.89		
10	*Alstonia spectabilis* R.Br.			0.71	0.72
11	*Anisoptera polyandra* Blume	15.14	13.25	2.59	3.11
12	*Anisoptera thurifera* Blume				2.88
13	*Anthocephalus cadamba* Miq.	3.05	4.34		1.07
14	*Aquilaria polyantha*			0.95	
15	*Arthocarpus* spp.	2.78	3.28		
16	*Baccaurea* sp			1.12	0.98
17	*Beilschmiedia* sp			9.26	7.27
18	*Blumeodendron amboinicum*			12.47	11.38
19	*Buchanania macrocarpa* Lauterb.			4.71	
20	*Calophyllum inophyllum* L.	6.63	3.84	1.57	0.78
21	*Calophyllum papuanum* Lauterb.	2.17	5.43	0.93	
22	*Campnosperma auriculatum* Hook.f.	1.62	1.98	3.38	0.76
23	*Cananga odorata* Hook. f. & Thomson		6.40		2.25
24	*Canarium asperum* Benth.	9.47	12.22	10.64	11.89

No.	Species Name	Bonggo Forest		Tunas Forest	
		4-year logged	8-year logged	4-year logged	8-year logged
		IVI (%)			
25	*Canarium australianum* F.Muell.	1.75		5.28	
26	*Canarium hirsutum* Willd.	1.33		2.78	3.93
27	*Canarium indicum* L.	8.13	5.16	8.85	9.68
28	*Canarium maluense* Lauterb.		1.49	2.36	
29	*Canarium* sp			2.72	
30	*Carallia brachiata* (Lour.) Merr.				1.16
31	*Celtis rigescens* Planch.	16.31	19.96		
32	*Cerbera floribunda* K. Schum.			0.92	
33	*Chisocheton ceramicus* C. DC.	3.46	5.55	6.61	4.17
34	*Cinnamomum culilawan* Blume		0.93		1.59
35	*Cleistanthus myrianthus* Kurz.			0.93	
36	*Crudia papuana* Kosterm.		0.62		
37	*Dacryodes* sp				2.87
38	*Dillenia alata* A. DC.		2.39		
39	*Dillenia ovalifolia* Hoogland.				0.70
40	*Diospyros* sp	4.51	3.80		
41	*Dracontomelum edule* Merr.	3.05	1.03		
42	*Drypetes globosa* Pax & Hoffm.			1.61	2.00
43	*Duabanga moluccana* Blume		0.69		
44	*Dysoxylum acutangulum* Miq.			1.16	
45	*Elaeocarpus sphaericus* Schum.		0.62		
46	*Endospermum medullosum* L.S.Sm.	1.52	2.87		
47	*Endospermum moluccanum* Beec.	4.51			
48	*Euodia bonwickii* F. Muell.				2.29
49	*Ficus benjamina* L.	3.99	5.24		
50	*Ficus glomerata* Roxb.			0.86	6.03

No.	Species Name	Bonggo Forest		Tunas Forest	
		4-year logged	8-year logged	4-year logged	8-year logged
		IVI (%)			
51	*Ficus nodosa* Teijsm. & Binn.				5.22
52	*Ficus variegata* Blume			2.24	1.09
53	*Ganophyllum falcatum* Blume				4.45
54	*Ganua boerlageana* Pierre			1.16	
55	*Garcinia celebica* L.			3.40	2.74
56	*Gironniera subaequalis* Plancho			8.11	4.52
57	*Gnetum gnemon* L.	1.64	1.58	0.88	1.64
58	*Gomphandra* sp			0.84	
59	*Haplolobus floribundus* H.J.Lam				1.60
60	*Homalium foetidum* Benth.	19.18	27.74		0.78
61	*Hopea papuana* Diels	1.48	1.58	5.24	6.64
62	*Intsia bijuga* Kuntze	3.14	6.49		
63	*Kayea floribunda* Wall				1.41
64	*Koordersiodendron papuanum*	1.26	1.74		
65	*Linociera macrophylla* Wall.			0.91	0.70
66	*Linociera oblongifolia* Koord.	0.98	1.50		
67	*Lithocarpus rufovillosus* Rehder			6.78	2.78
68	*Litsea firma*				3.48
69	*Litsea ledermannii* Teschner	15.27	15.65	4.29	1.92
70	*Litsea timoriana* Span.			13.19	14.26
71	*Litsea tuberculata* Boerl.			8.06	5.56
72	*Litsea irianensis* Kosterm.			0.90	2.68
73	*Macaranga beillei* Prain	5.59	1.53		
74	*Macaranga inermis* Pax & K.Hoffm.	2.74			1.14
75	*Macaranga mappa* (L.) Müll. Arg.				0.79
76	*Mangifera* sp	1.14	2.82		

No.	Species Name	Bonggo Forest		Tunas Forest	
		4-year logged	8-year logged	4-year logged	8-year logged
		IVI (%)			
77	*Manilkara fasciculata* Lam & Maas	1.85			
78	*Mastixiodendron pachyclados* Melch.	9.92	7.85		
79	*Mastixiodendron plectocarpum* S.P.Darwin			6.82	4.42
80	*Medusanthera polot*			7.16	7.20
81	*Myristica argentea*				4.50
82	*Myristica fatua* Houtt.	3.78	1.93	2.50	
83	*Myristica globosa* W.J. de Wilde			6.63	6.34
84	*Myristica hollrungii* Warb.				3.78
85	*Myristica lepidota* Blume	4.68	1.53	7.24	3.18
86	*Myristica neglecta* Warb.			2.20	0.80
87	*Myristica* sp	5.27	7.49		
88	*Myristica sulcata* Warb.	9.76	12.50	9.41	10.13
89	*Myristica tubiflora* Blume	10.27	9.93	8.84	6.41
90	*Neonauclea* sp			1.73	
91	*Neoscortechinia* sp			3.23	3.37
92	*Octomeles sumatrana* Miq.	3.82	5.92		
93	*Palaquium amboinense* Burck			1.12	
94	*Palaquium galactoxylum* H.J.Lam			2.23	3.65
95	*Palaquium lobbianum* Burck			1.16	0.68
96	*Palaquium obtusifolium* Burck	2.85	0.93	8.20	7.76
97	*Parartocarpus* spp	4.21	3.81		
98	*Pimelodendron amboinicum* Hassk.	11.53	7.93	10.59	9.38
99	*Planchonella anteridifera* H.J. Lam				0.61
100	*Planchonella firma* (Miq.) Dubard			0.91	6.25
101	*Podocarpus nerlifolius* D.Do		2.09		
102	*Pometia acuminata*	6.18	4.07	5.55	6.60

No.	Species Name	Bonggo Forest		Tunas Forest	
		4-year logged	8-year logged	4-year logged	8-year logged
		IVI (%)			
103	*Pometia pinnata* Forst. & Forst.	12.86	11.59	3.88	6.96
104	*Pterocarpus indicus* Willd.	8.75	4.59		
105	*Pterygota horsfieldii* Kosterm.	9.66	7.19		
106	*Saccopetalum* sp			0.89	2.20
107	*Semecarpus tannaensis* Guillaumin		0.85		
108	*Sloanea pullei* A.C. Sm.	0.98	0.63		
109	*Spondias dulcis* G. Forst.	2.91	1.76		
110	*Sterculia shillinglawii* F.Muell.			2.07	1.07
111	*Syzygium anomalum* Lauterb.	10.66	14.31	3.22	4.28
112	*Syzygium gustavioides* B.Hyland			0.96	
113	*Syzygium hylophilum* Merr. & Perry	5.66	4.30	7.94	1.32
114	*Syzygium* sp	4.63	4.49	0.86	
115	*Syzygium versteegii* Merr. & Perry	6.70	2.29	10.05	8.19
116	*Teijsmanniodendron bogoriense* Koord.		0.61	9.51	7.67
117	*Terminalia arborea* Koord.	9.25	2.45		
118	*Toona sureni* B. (Meril)	5.86	6.17		
119	*Urandra brassii* Howard				0.47
120	*Vatica rassak* Blume	0.84	3.42	22.16	26.34
121	*Vitex cofassus* Blume		1.06		
122	*Vitex pubescens* (L.) Vahl			4.66	1.31
123	*Weinmannia novoguineensis* Perry				1.05
124	*Whiteodendron papuana*			2.65	3.01

Curriculum vitae

Personal details

Name:	Yosias Gandhi
Nationality:	Indonesia
Date of Birth:	05.07.1966
Place of Birth:	Merauke, Indonesia
Sex:	Male
Marital status	Married

Education

2013-2021: Ph.D. student, Department of Tropical Silviculture and Forest Ecology, *Georg- August University of Göttingen, Germany*

1997-2000: M.Sc. in Wood Science and Forest Technology, *the University of the Philippines Los Banos, Laguna, Philippines*

1985-1991: Bachelor in Forestry, *University Cenderawasih, Manokwari, Indonesia*

Professional Experience

1995 - present: Lecturer in Faculty of Forestry, *the University of Papua, Manokwari, Indonesia*

Contact address

Faculty of Forestry
University of Papua
Jalan Gunung Salju Amban,
Manokwari 98314
INDONESIA
E-mail: y.gandhi@unipa.ac.id